# BIG IDEAS
# MATH®
## Modeling Real Life

## Grade 7
### Accelerated

## Student Journal

- Review & Refresh

- Exploration Journal

- Practice Worksheets

- Notetaking with Vocabulary

- Self-Assessments

- Exploration Manipulatives

**BIG IDEAS LEARNING®**

Erie, Pennsylvania

# About the Student Journal

### Review & Refresh

The Review & Refresh provides students the opportunity to practice prior skills necessary to move forward.

### Exploration Journal

The Exploration pages correspond to the Exploration in the Pupil Edition. Here students have room to show their work and record their answers.

### Practice Worksheets

Each section of the Pupil Edition has an additional practice on the key concepts taught in the lesson.

### Notetaking with Vocabulary

The student-friendly notetaking component is designed to be a reference for key vocabulary from the lesson. There is room for students to add definition to their words and take notes about key ideas.

### Self-Assessment

For every lesson, students can rate their understanding of the learning target and success criteria.

### Exploration Manipulatives

Manipulatives needed for the explorations are included in the back of the Student Journal.

ISBN 13: 978-1-64245-370-6

3456789-22 21 20 19

# Contents

# Contents

# Contents

# Contents

# Contents

# Contents

# Contents

# Contents

# Contents

**Big Ideas Math: Modeling Real Life Grade 7 Accelerated**
Student Journal

**xi**

# Contents

# Contents

# Contents

## Chapter 1  Review & Refresh

**Simplify the expression. Explain each step.**

**1.** $2 + (5 + y)$

**2.** $(c + 1) + 9$

**3.** $(2.3 + n) + 1.4$

**4.** $7 + (d + 5)$

**5.** $(t + 3) + 0$

**6.** $0 + (g + 4)$

## Chapter 1  Review & Refresh (continued)

**Evaluate the expression.**

7. $\dfrac{1}{8} + \dfrac{1}{9}$

8. $\dfrac{2}{3} + \dfrac{9}{10}$

9. $\dfrac{7}{12} - \dfrac{1}{4}$

10. $\dfrac{6}{7} - \dfrac{4}{5}$

11. You have 8 cups of sugar. A recipe calls for $\dfrac{2}{3}$ cup of sugar. Another recipe calls for $\dfrac{1}{4}$ cup of sugar. How much sugar do you have left after making the recipes?

## 1.1 Rational Numbers
**For use with Exploration 1.1**

**Learning Target:** Understand absolute values and ordering of rational numbers.

**Success Criteria:**
- I can graph rational numbers on a number line.
- I can find the absolute value of a rational number.
- I can use a number line to compare rational numbers.

---

**1 EXPLORATION: Using a Number Line**

**Work with a partner. Make a number line on the floor. Include both negative numbers and positive numbers.**

a. Stand on an integer. Then have your partner stand on the opposite of the integer. How far are each of you from 0? What do you call the distance between a number and 0 on a number line?

b. Stand on a rational number that is not an integer. Then have your partner stand on any other number. Which number is greater? How do you know?

---

**Big Ideas Math: Modeling Real Life Grade 7 Accelerated**
Student Journal

## 1.1 Rational Numbers (continued)

c. Stand on any number other than 0 on the number line. Can your partner stand on a number that is:

greater than your number and farther from 0?

greater than your number and closer to 0?

less than your number and the same distance from 0?

less than your number and farther from 0?

For each case in which it is not possible to stand on a number as directed, explain why it is not possible. In each of the other cases, how can you decide where your partner can stand?

## 1.1 Notetaking with Vocabulary

**Vocabulary:**

**Notes:**

## 1.1 Self-Assessment

**Use the scale below to rate your understanding of the learning target and the success criteria.**

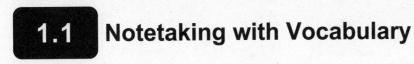

| 1 | 2 | 3 | 4 |
|---|---|---|---|
| I do not understand. | I can do it with help. | I can do it on my own. | I can teach someone else. |

| | Rating | Date |
|---|---|---|
| **1.1 Rational Numbers** | | |
| **Learning Target**: Understand absolute values and ordering of rational numbers. | 1   2   3   4 | |
| I can graph rational numbers on a number line. | 1   2   3   4 | |
| I can find the absolute value of a rational number. | 1   2   3   4 | |
| I can use a number line to compare rational numbers. | 1   2   3   4 | |

# 1.1 Practice

**Complete the statement using <, >, or =.**

1. $|-23|$ ____ 23          2. $|-142|$ ____ $|-157|$          3. $-|-78|$ ____ 52

4. You and your friend are swimming against the current. You move forward 15 feet. Your friend is not a strong swimmer, so he moves back 6 feet.

   a. Write each amount as an integer.

   b. Write each amount as the distance swam.

**Order the values from least to greatest.**

5. $14, |-25|, -|-34|, 28, |0|$          6. $|-16|, 10, |25|, -16, |-43|$

7. The boiling point of a liquid is the temperature at which the vapor pressure of the liquid equals the environmental pressure surrounding the liquid.

| Substance | Hydrogen | Oxygen | Iodine | Phosphorus |
|---|---|---|---|---|
| Boiling Point (°C) | -253 | -183 | 184 | 280 |

   a. Which substance in the table has the highest boiling point? Explain.

   b. Is the boiling point of oxygen or iodine closer to 0°C? Explain.

8. You are riding a rollercoaster.

   a. Your velocity is 13 feet per second. Are you moving up or moving down?

   b. What is your speed in part (a)? Give the units.

   c. Your velocity is −17 feet per second. Are you moving up or moving down?

   d. What is your speed in part (c)? Give the units.

9. There is one integer for which there does not exist another integer with the same absolute value. What is that integer?

**Determine whether the statement is *true* or *false*. Explain your reasoning.**

10. The absolute value of 3 above par is the same as the absolute value of 3 below par.

11. If $x < 0$, then $|x| < x$.

## 1.2  Adding Integers
### For use with Exploration 1.2

**Learning Target:**    Find sums of integers.

**Success Criteria:**    • I can explain how to model addition of integers on a number line.
• I can find sums of integers by reasoning about absolute values.
• I can explain why the sum of a number and its opposite is 0.

---

**1**  **EXPLORATION:** Using Integer Counters to Find Sums

**Work with a partner. You can use the integer counters shown at the right to find sums of integers.**

$$\boxed{+} = +1$$
$$\boxed{-} = -1$$

a.  How can you use integer counters to model a sum?
a sum that equals 0?

b.  What expression is being modeled below? What is the value of the sum?

## 1.2 Adding Integers (continued)

**c.** Use integer counters to complete the table.

| Expression | Type of Sum | Sum | Sum: Positive, Negative, or Zero |
|---|---|---|---|
| $-3 + 2$ | Integers with different sign | | |
| $-4 + (-3)$ | | | |
| $5 + (-3)$ | | | |
| $7 + (-7)$ | | | |
| $2 + 4$ | | | |
| $-6 + (-2)$ | | | |
| $-5 + 9$ | | | |
| $15 + (-9)$ | | | |
| $-10 + 10$ | | | |
| $-6 + (-6)$ | | | |
| $13 + (-13)$ | | | |

**d.** How can you tell whether the sum of two integers is *positive*, *negative*, or *zero*?

**e.** Write rules for adding (i) two integers with the same sign, (ii) two integers with different signs, and (iii) two opposite integers.

**Notetaking with Vocabulary**

**Vocabulary:**

**Notes:**

**1.2** **Self-Assessment**

Use the scale below to rate your understanding of the learning target and the success criteria.

| 1 | 2 | 3 | 4 |
|---|---|---|---|
| I do not understand. | I can do it with help. | I can do it on my own. | I can teach someone else. |

| | Rating | Date |
|---|---|---|
| **1.2 Adding Integers** | | |
| **Learning Target**: Find sums of integers. | 1  2  3  4 | |
| I can explain how to model addition of integers on a number line. | 1  2  3  4 | |
| I can find sums of integers by reasoning about absolute values. | 1  2  3  4 | |
| I can explain why the sum of a number and its opposite is 0. | 1  2  3  4 | |

## 1.2 Practice

1. The elevation of your plot of land is 2 feet below sea level. You add 7 feet of dirt to your land. What is the new elevation of your land?

**Tell how the Commutative and Associative Properties of Addition can help you find the sum using mental math. Then find the sum.**

2. $18 + (-25) + (-18)$    3. $-22 + 45 + (-8)$    4. $28 + (-12) + 4$

**Find the sum.**

5. $(-63) + 81 + 0$        6. $101 + (-51) + (-36)$   7. $(-117) + 125 + (-67)$

**Describe the location of the sum, relative to *p*, on a number line.**

8. $p + 5$                 9. $p + (-2)$              10. $p + (-q)$

**Use mental math to solve the equation.**

11. $n + (-20) = 5$        12. $c + (-71) = 0$         13. $-30 + k = -110$

14. Write three integers that do not all have the same sign that have a sum of –20. Write three integers that do not all have the same sign that have a sum of 10.

15. The temperature at 6 A.M. is –12°F. During the next twelve hours, the temperature increases 25°F. During the following 5 hours, the temperature decreases 23°F. What is the temperature at 11 P.M.?

16. Complete the magic square so that each row and column has a magic sum of 0. Use integers from –9 to 9, without repeating an integer.

| 9 |  | –3 |
|---|---|---|
|  |  |  |
|  | 2 |  |

17. Consider the integers *p* and *q*. Describe all of the possible values of *p* and *q* for each circumstance. Justify your answers.

a. $|p| + |-q| = 0$       b. $|p| + |-q| < 0$       c. $|p| + |-q| > 0$

## 1.3 Adding Rational Numbers
**For use with Exploration 1.3**

**Learning Target:** Find sums of rational numbers.

**Success Criteria:**
- I can explain how to model addition of rational numbers on a number line.
- I can find sums of rational numbers by reasoning about absolute values.
- I can use properties of addition to efficiently add rational numbers.

---

**1** **EXPLORATION:** Adding Rational Numbers

**Work with a partner.**

a. Choose a unit fraction to represent the space between the tick marks on each number line. What addition expressions are being modeled? What are the sums?

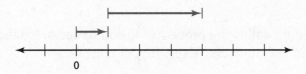

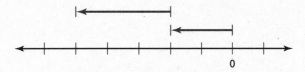

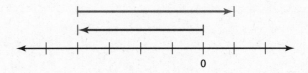

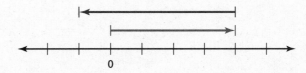

**1.3** **Adding Rational Numbers** (continued)

**b.** Do the rules for adding integers apply to all rational numbers? Explain your reasoning.

**c.** You have used the following properties to add integers. Do these properties apply to all rational numbers? Explain your reasoning.

Commutative Property of Addition

Associative Property of Addition

Additive Inverse Property

**Notetaking with Vocabulary**

**Vocabulary:**

**Notes:**

**1.3** **Self-Assessment**

Use the scale below to rate your understanding of the learning target and the success criteria.

| **1** | **2** | **3** | **4** |
|---|---|---|---|
| I do not understand. | I can do it with help. | I can do it on my own. | I can teach someone else. |

|  | Rating | Date |
|---|---|---|
| **1.3 Adding Rational Numbers** | | |
| **Learning Target**: Find sums of rational numbers. | 1  2  3  4 | |
| I can explain how to model addition of rational numbers on a number line. | 1  2  3  4 | |
| I can find sums of rational numbers by reasoning about absolute values. | 1  2  3  4 | |
| I can use properties of addition to efficiently add rational numbers. | 1  2  3  4 | |

## 1.3 Practice

1. Describe and correct the error in finding the sum.

$$\times \quad 2\frac{5}{6} + \left(-\frac{8}{15}\right) = \frac{13}{6} + \left(-\frac{8}{15}\right) = \frac{65 + (-16)}{30} = \frac{49}{30} = 1\frac{19}{30}$$

**Evaluate the expression when** $x = -\frac{1}{5}$ **and** $y = \frac{3}{4}$**.**

2. $x + (-y)$

3. $4x + y$

4. $-|x| + y$

5. Your banking account is –$1.56. You deposit $10. What is your new balance?

6. You mow $\frac{1}{3}$ of the lawn. Your sister mows $\frac{2}{7}$ of the lawn. What fraction of the lawn is not mowed?

**Find the sum. Write fractions in simplest form.**

7. $1\frac{1}{4} + \left(-4\frac{1}{5}\right) + \left(-2\frac{3}{5}\right)$

8. $-\frac{1}{3} + 2\frac{2}{9} + \left(-5\frac{2}{3}\right)$

9. $-1.5 + (14.2) + 7.3$

10. When is the sum of two rational numbers with different signs positive?

11. The table at the right shows the amount of snowfall (in inches) for three months compared to the yearly average. Is the snowfall for the three-month period greater than or less than the yearly average? Explain.

| December | January | February |
|----------|---------|----------|
| $1\frac{2}{3}$ | $-2\frac{1}{6}$ | $2\frac{5}{6}$ |

12. The table below shows the weekly profits of a concession stand. What must the Week 5 profit be to break even over the 5-month period?

| Week 1 | Week 2 | Week 3 | Week 4 | Week 5 |
|--------|--------|--------|--------|--------|
| 2.4 | –1.7 | 5.4 | –3.75 | |

13. When is the sum of two positive decimal numbers an integer?

## 1.4 Subtracting Integers
### For use with Exploration 1.4

**Learning Target:** Find differences of integers.

**Success Criteria:**
- I can explain how subtracting integers is related to adding integers.
- I can explain how to model subtraction of integers on a number line.
- I can find differences of integers by reasoning about absolute values.

**1 EXPLORATION: Using Integer Counters to Find Differences**

**Work with a partner.**

**a.** Use integer counters to find the following sum and difference. What do you notice?

$$4 + (-2) \qquad 4 - 2$$

| + | = +1 |
| - | = -1 |

**b.** In part (a), you *removed* zero pairs to find the sums. How can you use integer counters and zero pairs to find $-3 - 1$?

**1.4** **Subtracting Integers** (continued)

c. Use integer counters to complete the table.

| Expression | Operation: Add or Subtract | Answer |
|---|---|---|
| 4 – 2 | Subtract 2 | |
| 4 + (–2) | | |
| –3 – 1 | | |
| –3 + (–1) | | |
| 3 – 8 | | |
| 3 + (–8) | | |
| 9 – 13 | | |
| 9 + (–13) | | |
| –6 – (–3) | | |
| –6 + 3 | | |
| –5 – (–12) | | |
| –5 + 12 | | |

d. Write a general rule for subtracting integers.

Name_____ Date_____

## 1.4 Notetaking with Vocabulary

**Vocabulary:**

**Notes:**

## 1.4 Self-Assessment

Use the scale below to rate your understanding of the learning target and the success criteria.

| 1 | 2 | 3 | 4 |
|---|---|---|---|
| I do not understand. | I can do it with help. | I can do it on my own. | I can teach someone else. |

|  | Rating | Date |
|---|---|---|
| **1.4 Subtracting Integers** | | |
| **Learning Target**: Find differences of integers. | 1  2  3  4 | |
| I can explain how subtracting integers is related to adding integers. | 1  2  3  4 | |
| I can explain how to model subtraction of integers on a number line. | 1  2  3  4 | |
| I can find differences of integers by reasoning about absolute values. | 1  2  3  4 | |

# 1.4 Practice

1. A dolphin is at −28 feet. It swims up and jumps out of the water to a height of 8 feet. Find the vertical distance the dolphin travels. Justify your answer.

**Evaluate the expression.**

2. $0 - (-41) - 28$

3. $-51 - (-23) + (-16)$

4. $8 - (-103) - (-95)$

**Tell how the Commutative and Associative Properties of Addition can help you evaluate the expression using mental math. Then evaluate the expression.**

5. $-(-22) + 17 - 22$

6. $[-15 + (-31)] + 15$

7. $19 + (-19 - 24)$

8. The table shows the record monthly high and low temperatures in International Falls, Minnesota.

| | Jan | Feb | Mar | Apr | May | Jun | Jul | Aug | Sep | Oct | Nov | Dec |
|---|---|---|---|---|---|---|---|---|---|---|---|---|
| **High (°F)** | 48 | 58 | 76 | 93 | 95 | 99 | 98 | 95 | 95 | 88 | 73 | 57 |
| **Low (°F)** | −46 | −45 | −38 | −14 | 11 | 23 | 34 | 30 | 20 | 2 | −32 | −41 |

   a. Find the range of temperatures for each month.

   b. What are the all-time high and all-time low temperatures?

   c. What is the range of the temperatures in part (b)?

**Use mental math to solve the equation.**

9. $c - (-15) = 38$

10. $-5 - k = -12$

11. $-25 - m = 28$

12. For what values of $a$ and $b$ is the statement $|a - b| = |a| - |b|$ false?

13. Write two different pairs of negative integers, $x$ and $y$, that make the statement $x - y = 6$ true.

**Describe the location of the difference, relative to $p$, on a number line.**

14. $p - 3$

15. $p - (-5)$

16. $p - q$

17. Write two different triples of integers, $x$, $y$, and $z$, that meet the following conditions.

   a. They do not all have the same sign and $x - y - z = 0$.

   b. They all have the same sign and $(-x) + y - z = -17$.

# 1.5 Subtracting Rational Numbers
**For use with Exploration 1.5**

**Learning Target:** Find differences of rational numbers and find distances between numbers on a number line.

**Success Criteria:**
- I can explain how to model subtraction of rational numbers on a number line.
- I can find differences of rational numbers by reasoning about absolute values.
- I can find distances between numbers on a number line.

## 1 ) EXPLORATION: Subtracting Rational Numbers

**Work with a partner.**

a. Choose a unit fraction to represent the space between the tick marks on each number line. What expressions involving subtraction are being modeled? What are the differences.?

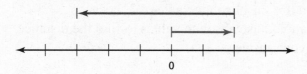

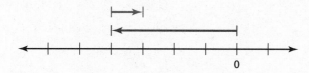

b. Do the rules for subtracting integers apply to all rational numbers? Explain your reasoning?

c. You have used the commutative and associative properties to add integers. Do these properties apply in expressions involving subtraction? Explain your reasoning.

## 1.5 Subtracting Rational Numbers (continued)

**2 EXPLORATION:** Finding Distances on a Number Line

**Work with a partner.**

**a.** Find the distance between 3 and –2 on a number line.

**b.** The distance between 3 and 0 is the absolute value of 3, because $|3 - 0| = |3| = 3$. How can you use absolute values to find the distance between 3 and –2? Justify your answer.

**c.** Choose any two rational numbers. Use your method in part (b) to find the distance between the numbers. Use a number line to check your answer.

## 1.5 Notetaking with Vocabulary

**Vocabulary:**

**Notes:**

## 1.5 Self-Assessment

**Use the scale below to rate your understanding of the learning target and the success criteria.**

| 1 | 2 | 3 | 4 |
|---|---|---|---|
| I do not understand. | I can do it with help. | I can do it on my own. | I can teach someone else. |

|  | Rating | Date |
|---|---|---|
| **1.5 Subtracting Rational Numbers** | | |
| **Learning Target:** Find differences of rational numbers and find distances between numbers on a number line. | 1  2  3  4 | |
| I can explain how to model subtraction of rational numbers on a number line. | 1  2  3  4 | |
| I can find differences of rational numbers by reasoning about absolute values. | 1  2  3  4 | |
| I can find distances between numbers on a number line. | 1  2  3  4 | |

# 1.5 Practice

**Find the distance between the two numbers on a number line.**

1. $-7\frac{1}{5}, -4\frac{2}{3}$

2. $-9.2, 4.5$

3. $-2, -3.7$

4. The largest orange in a bag has a circumference of $9\frac{5}{8}$ inches. The smallest orange has a circumference of $7\frac{13}{16}$ inches. Write the difference of the circumferences of the smallest orange and the largest orange. Then find the difference.

**Evaluate the expression.**

5. $\frac{5}{12} - \left(-3\frac{1}{4}\right) + \left(-6\frac{1}{2}\right) - 3$

6. $23.706 - (-82.31) - 130.641$

7. $-\frac{3}{8} - (-4.35)$

8. $-\frac{5}{18} - \left|-\frac{1}{6}\right| + \left(-\frac{7}{9}\right)$

9. Your bank account balance is $32.00. You make the following withdrawals, in the following order: $15.00, $7.41, $35.79, and $0.53. After each withdrawal that leaves a negative balance, the bank adds a −$32.00 bank fee to your account. What is your new balance?

10. Fill in the blanks to make the solution correct.

$$3\frac{3}{4} - \boxed{\phantom{x}}\frac{\boxed{\phantom{x}}}{8} = 2$$

11. When is the difference of two mixed numbers an integer? Explain.

12. Points $A\left(5, \frac{2}{3}\right)$ and $B\left(5, 1\frac{1}{6}\right)$ lie in a coordinate plane.

   a. Find the distance between the points based on the $x$-coordinates. Interpret your answer.

   b. Find the distance between the points based on the $y$-coordinates.

   c. Write the expression for the difference of the $y$-coordinates, relative to point $A$.

   d. Write the expression for the difference of the $y$-coordinates, relative to point $B$.

   e. If the $y$-coordinate represents the number of bushels of apples picked, which of your answers, *part (c)* or *part (d)*, agrees with the statement "The number of bushels of apples picked is $\frac{1}{2}$ less than last year."?

Name_____ Date_____ 23

Use the scale below to rate your understanding of the learning target and the success criteria.

| **1** | **2** | **3** | **4** |
|---|---|---|---|
| I do not understand. | I can do it with help. | I can do it on my own. | I can teach someone else. |

|  | Rating | Date |
|---|---|---|
| **1.1 Rational Numbers** | | |
| **Learning Target:** Understand absolute values and ordering of rational numbers. | 1  2  3  4 | |
| I can graph rational numbers on a number line. | 1  2  3  4 | |
| I can find the absolute value of a rational number. | 1  2  3  4 | |
| I can use a number line to compare rational numbers. | 1  2  3  4 | |
| **1.2 Adding Integers** | | |
| **Learning Target:** Find sums of integers. | 1  2  3  4 | |
| I can explain how to model addition of integers on a number line. | 1  2  3  4 | |
| I can find sums of integers by reasoning about absolute values. | 1  2  3  4 | |
| I can explain why the sum of a number and its opposite is 0. | 1  2  3  4 | |
| **1.3 Adding Rational Numbers** | | |
| **Learning Target:** Find sums of rational numbers. | 1  2  3  4 | |
| I can explain how to model addition of rational numbers on a number line. | 1  2  3  4 | |
| I can find sums of rational numbers by reasoning about absolute values. | 1  2  3  4 | |
| I can use properties of addition to efficiently add rational numbers. | 1  2  3  4 | |

**Chapter 1**

# Chapter Self-Assessment (continued)

|  | Rating | Date |
|---|---|---|
| **1.4 Subtracting Integers** | | |
| **Learning Target:** Find differences of integers. | 1   2   3   4 | |
| I can explain how subtracting integers is related to adding integers. | 1   2   3   4 | |
| I can explain how to model subtraction of integers on a number line. | 1   2   3   4 | |
| I can find differences of integers by reasoning about absolute values. | 1   2   3   4 | |
| **1.5 Subtracting Rational Numbers** | | |
| **Learning Target:** Find differences of rational numbers and find distances between numbers on a number line. | 1   2   3   4 | |
| I can explain how to model subtraction of rational numbers on a number line. | 1   2   3   4 | |
| I can find differences of rational numbers by reasoning about absolute values. | 1   2   3   4 | |
| I can find distances between numbers on a number line. | 1   2   3   4 | |

## Chapter 2 Review & Refresh

**Simplify the expression. Explain each step.**

**1.** $10(7t)$

**2.** $8(4k)$

**3.** $13 \cdot 0 \cdot p$

**4.** $7 \cdot z \cdot 0$

**5.** $2.5 \cdot w \cdot 1$

**6.** $1 \cdot x \cdot 19$

**Write the decimal as a fraction.**

**7.** $0.26$

**8.** $0.79$

**Write the fraction as a decimal.**

**9.** $\frac{4}{10}$

**10.** $\frac{17}{20}$

**11.** A quarterback completed 0.6 of his passes during a game. Write the decimal as a fraction.

**Write the mixed number as an improper fraction.**

**12.** $-3\frac{1}{5}$

**13.** $4\frac{3}{7}$

**14.** $8\frac{2}{3}$

**15.** $-10\frac{5}{6}$

**Evaluate the expression.**

**16.** $\frac{5}{9} \cdot \frac{1}{3}$

**17.** $\frac{8}{15} \cdot \frac{3}{4}$

**18.** $\frac{7}{8} \div \frac{11}{16}$

**19.** $\frac{3}{10} \div \frac{2}{5}$

## 2.1 Multiplying Integers
**For use with Exploration 2.1**

**Learning Target:**  Find products of integers.

**Success Criteria:**  • I can explain the rules for multiplying integers.

• I can find products of integers with the same sign.

• I can find products of integers with different signs.

---

**1  EXPLORATION:** Understanding Products Involving Negative Integers

**Work with a partner.**

**a.** The number line and integer counters model the product $3 \cdot 2$. How can you find $3 \cdot (-2)$? Explain.

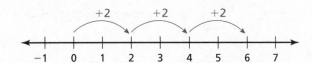

**b.** Use the tables to find $-3 \cdot 2$ and $-3 \cdot (-2)$. Explain your reasoning.

| |
|---|
| $2 \cdot 2 = 4$ |
| $1 \cdot 2 = 2$ |
| $0 \cdot 2 = 0$ |
| $-1 \cdot 2 = $ ___ |
| $-2 \cdot 2 = $ ___ |
| $-3 \cdot 2 = $ ___ |

| |
|---|
| $-3 \cdot 3 = -9$ |
| $-3 \cdot 2 = -6$ |
| $-3 \cdot 1 = -3$ |
| $-3 \cdot 0 = $ ___ |
| $-3 \cdot -1 = $ ___ |
| $-3 \cdot -2 = $ ___ |

---

## 2.1 Multiplying Integers (continued)

**c.** Complete the table. Then write general rules for multiplying (i) two integers with the same sign and (ii) two integers with different signs.

| Expression | Type of Product | Product | Product: Positive or Negative |
|---|---|---|---|
| $3 \cdot 2$ | Integers with the same sign | | |
| $3 \cdot (-2)$ | | | |
| $-3 \cdot 2$ | | | |
| $-3 \cdot (-2)$ | | | |
| $6 \cdot 3$ | | | |
| $2 \cdot (-5)$ | | | |
| $-6 \cdot 5$ | | | |
| $-5 \cdot (-3)$ | | | |

## 2.1 Notetaking with Vocabulary

**Vocabulary:**

**Notes:**

## 2.1 Self-Assessment

**Use the scale below to rate your understanding of the learning target and the success criteria.**

| 1 | 2 | 3 | 4 |
|---|---|---|---|
| I do not understand. | I can do it with help. | I can do it on my own. | I can teach someone else. |

|  | Rating | Date |
|---|---|---|
| **2.1 Multiplying Integers** | | |
| **Learning Target:** Find products of integers. | 1   2   3   4 | |
| I can explain the rules for multiplying integers. | 1   2   3   4 | |
| I can find products of integers with the same sign. | 1   2   3   4 | |
| I can find products of integers with different signs. | 1   2   3   4 | |

Name _____ Date _____

1. The water in a pool evaporates at a rate of 16 gallons per week. What integer represents the change in the number of gallons of water in the pool after 24 weeks?

**Evaluate the expression.**

2. $-12^2$

3. $(-2)^3 \cdot (-3)^2$

4. $-9 \cdot 0 \cdot (-3)$

5. $-|-3| \cdot (-6)$

6. $11(-3) - (-2)(7)$

7. $-5 \cdot 8 - (-4)^3$

8. The gym offers a discount when more than one member of the family joins. The first member ($n = 0$) pays \$550 per year. The second member to join ($n = 1$) gets a discount of \$75 per year. The third member ($n = 2$) gets an additional \$75 discount. The price for the $n$th member is given by $550 + (-75n)$.

   a. What is the price for the fourth member to join ($n = 3$)?

   b. For a large family, is it possible that a member would join for free? If so, which member would it be? Explain your reasoning.

   c. Other than \$0, what is the lowest amount that a member would pay to join? Which member would it be? Explain your reasoning.

9. Two integers, $a$ and $b$, have a product of $-48$.

   a. What is the greatest possible sum of $a$ and $b$?

   b. Is it possible for $a$ and $b$ to have a sum of 13? If so, what are the integers?

   c. What is the least possible difference of $a$ and $b$?

**Tell whether the statement is *true* or *false*. Explain your reasoning.**

10. The product of three negative integers is negative.

11. The product of four negative integers is negative.

12. The product of two negative integers and one positive integer is negative.

13. The product of one negative integer and two positive integers is negative.

14. If there is an odd number of negative integers, then the product will be negative.

## 2.2 Dividing Integers
**For use with Exploration 2.2**

**Learning Target:** Find quotients of integers.

**Success Criteria:**
- I can explain the rules for dividing integers.
- I can find quotients of integers with the same sign.
- I can find quotients of integers with different signs.

**1   EXPLORATION: Understanding Quotients Involving Negative Integers**

**Work with a partner.**

a. Discuss the relationship between multiplication and division with your partner.

b. Complete the table. Then write general rules for dividing (i) two integers with the same sign and (ii) two integers with different signs.

| Expression | Type of Quotient | Quotient | Quotient: Positive, Negative, or Zero |
|---|---|---|---|
| $-15 \div 3$ | Integers with different signs | | |
| $12 \div (-6)$ | | | |
| $10 \div (-2)$ | | | |
| $-6 \div 2$ | | | |
| $-12 \div (-12)$ | | | |
| $-21 \div (-7)$ | | | |
| $0 \div (-15)$ | | | |
| $0 \div 4$ | | | |
| $-5 \div 4$ | | | |
| $5 \div (-4)$ | | | |

**2.2** **Dividing Integers** (continued)

**c.** Find the values of $-\frac{8}{4}$, $\frac{-8}{4}$, and $\frac{8}{-4}$. What do you notice? Is it true for $-\frac{a}{b}$, $\frac{-a}{b}$, and $\frac{a}{-b}$ when $a$ and $b$ are integers? Explain.

**d.** Is every quotient of integers a rational number? Explain your reasoning.

**Notetaking with Vocabulary**

**Vocabulary:**

**Notes:**

 **Self-Assessment**

Use the scale below to rate your understanding of the learning target and the success criteria.

| 1 | 2 | 3 | 4 |
|---|---|---|---|
| I do not understand. | I can do it with help. | I can do it on my own. | I can teach someone else. |

| | Rating | Date |
|---|:---:|:---:|
| **2.2 Dividing Integers** | | |
| **Learning Target**: Find quotients of integers. | 1  2  3  4 | |
| I can explain the rules for dividing integers. | 1  2  3  4 | |
| I can find quotients of integers with the same sign. | 1  2  3  4 | |
| I can find quotients of integers with different signs. | 1  2  3  4 | |

## 2.2 Practice

**Find the quotient, if possible.**

1. $\dfrac{-144}{-9}$     2. $\dfrac{0}{25}$     3. $-15 \div (-15)$     4. $-82 \div 0$

5. $-96 \div 8$     6. $225 \div (-25)$     7. $\dfrac{-156}{3}$     8. $\dfrac{99}{-9}$

9. Your team catches 42 Mahi Mahi over 2 weeks. What is the average daily Mahi Mahi catch?

**Evaluate the expression.**

10. $(-68) \div (-4) + 5 \cdot (-3)$     11. $10 - 12^2 \div (-2)^3$

12. $-|-16| \div (-8) \cdot 5^2$     13. $\dfrac{3 + 7 \cdot (-3^2)}{-5}$

14. PI-Squared and Euler Circles are in a math competition consisting of 10 two-part questions. Both parts correct earns 5 points, one part correct earns 2 points, and no parts correct earns –1 point.

   a. What is the mean points per question for PI-Squared?

   b. What is the mean points per question for Euler Circle?

   | Team | Both | One | None |
   |------|------|-----|------|
   | PI-Squared | 4 | 2 | 4 |
   | Euler Circles | 2 | 6 | 2 |

   c. Which team should win the competition? Explain your reasoning.

15. A 155-pound person burns about 500 calories per hour playing racquetball.

   a. One pound is equal to 3500 calories. How long will it take to burn 1 pound playing racquetball?

   b. How long will it take to burn 5 pounds playing racquetball? Explain your reasoning.

   c. If the person were to rest 5 minutes every 30 minutes of playing, how long would it take to burn 1 pound?

16. Find the next two numbers in the pattern $729, -243, 81, -27, \ldots$. Explain your reasoning.

## 2.3 Converting Between Fractions and Decimals
**For use with Exploration 2.3**

**Learning Target:** Convert between different forms of rational numbers.

**Success Criteria:**
- I can explain the difference between terminating and repeating decimals.
- I can write fractions and mixed numbers as decimals.
- I can write decimals as fractions and mixed numbers.

---

**1 EXPLORATION:** Analyzing Denominators of Decimal Fractions

**Work with a partner.**

**a.** Write each decimal as a fraction or mixed number.

0.7          1.29          12.831          0.0041

**b.** What do the factors of the denominators of the fractions you wrote have in common? Is this always true for decimal fractions?

---

**Big Ideas Math: Modeling Real Life Grade 7 Accelerated**

**2.3** **Converting Between Fractions and Decimals** (continued)

**2** **EXPLORATION:** Exploring Decimal Representations

**Work with a partner.**

**a.** A fraction $\frac{a}{b}$ can be interpreted as $a \div b$. Use a calculator to convert each unit fraction to a decimal. Do some of the decimals look different than the others? Explain.

$$\frac{1}{2} \qquad \frac{1}{3} \qquad \frac{1}{4} \qquad \frac{1}{5}$$

$$\frac{1}{6} \qquad \frac{1}{7} \qquad \frac{1}{8}$$

$$\frac{1}{9} \qquad \frac{1}{10} \qquad \frac{1}{11} \qquad \frac{1}{12}$$

**b.** Compare and contrast the fractions in part (s) with the fractions you wrote in Exploration 1. What conclusions can you make?

**c.** Does every fraction have a decimal form that either *terminates* or *repeats*? Explain your reasoning.

**Notetaking with Vocabulary**

**Vocabulary:**

**Notes:**

## 2.3 Self-Assessment

Use the scale below to rate your understanding of the learning target and the success criteria.

| 1 | 2 | 3 | 4 |
|---|---|---|---|
| I do not understand. | I can do it with help. | I can do it on my own. | I can teach someone else. |

|  | Rating | Date |
|---|---|---|
| **2.3 Converting Between Fractions and Decimals** | | |
| **Learning Target:** Convert between different forms of rational numbers. | 1 2 3 4 | |
| I can explain the difference between terminating and repeating decimals. | 1 2 3 4 | |
| I can write fractions and mixed numbers as decimals. | 1 2 3 4 | |
| I can write decimals as fractions and mixed numbers. | 1 2 3 4 | |

## 2.3 Practice

**Write the fraction or mixed number as a decimal.**

**1.** $-5\frac{3}{40}$

**2.** $-\frac{3}{22}$

**3.** $4\frac{1}{15}$

**4.** $-9\frac{1}{9}$

**Write the decimal as a fraction or mixed number in simplest form.**

**5.** $0.68$

**6.** $8.745$

**7.** $-9.98$

**8.** $-10.452$

**9.** You caught a red snapper that is $8\frac{5}{12}$ inches long. Your friend caught a red snapper that is $8\frac{6}{13}$ inches long. Who caught the larger red snapper?

**Complete the statement using <, >, or =.**

**10.** $0.13$ ____ $\frac{1}{8}$

**11.** $-1\frac{2}{9}$ ____ $-\frac{5}{4}$

**12.** $-5.175$ ____ $-5\frac{1}{6}$

**13.** Find one terminating and one repeating decimal between $-1\frac{1}{2}$ and $-1\frac{7}{9}$.

**14.** The table gives the tidal changes in the water level of a lagoon for every six hours of a given day.

| Time | 4:00 A.M. | 10:00 A.M. | 4:00 P.M. | 10:00 P.M. |
|---|---|---|---|---|
| Change (feet) | 2.25 | $-2\frac{6}{7}$ | $-\frac{3}{2}$ | $2\frac{1}{3}$ |

**a.** Order the numbers from least to greatest.

**b.** At what time(s) did the water level decrease?

**c.** What was the largest change in water level?

**d.** Did the tidal change in part (c) involve an increase or a decrease in water level?

**e.** Will the next tidal change be an increase or decrease in water level? Explain.

**15.** Let $a$ and $b$ be integers.

**a.** When can $\frac{a}{b}$ be written as a negative, repeating decimal between 0 and $-1$?

**b.** When can $\frac{a}{b}$ be written as a negative, terminating decimal between 0 and $-1$?

**c.** When can $\frac{a}{b}$ be written as a negative decimal between 0 and $-1$ that does not repeat and does not terminate?

## 2.4 Multiplying Rational Numbers
### For use with Exploration 2.4

**Learning Target:** Find products of rational numbers.

**Success Criteria:**
- I can explain the rules for multiplying rational numbers.
- I can find products of rational numbers with the same sign.
- I can find products of rational numbers with different signs.

**1 EXPLORATION:** Finding Products of Rational Numbers

**Work with a partner.**

a. Write a multiplication expression represented by each area model. Then find the product.

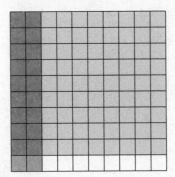 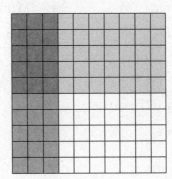

**2.4** **Multiplying Rational Numbers** (continued)

    **b.** Complete the table.

| | Expression | Product | Expression | Product |
|---|---|---|---|---|
| **i.** | $0.2 \times 0.9$ | | $-0.2 \times 0.9$ | |
| **ii.** | $0.3(0.5)$ | | $0.3(-0.5)$ | |
| **iii.** | $\dfrac{1}{4} \cdot \dfrac{1}{2}$ | | $\dfrac{1}{4} \cdot \left(-\dfrac{1}{2}\right)$ | |
| **iv.** | $1.2(0.4)$ | | $-1.2(-0.4)$ | |
| **v.** | $\dfrac{3}{10}\left(\dfrac{2}{5}\right)$ | | $-\dfrac{3}{10}\left(-\dfrac{2}{5}\right)$ | |
| **vi.** | $0.6 \times 1.8$ | | $-0.6 \times 1.8$ | |
| **vii.** | $1\dfrac{1}{4} \cdot 2\dfrac{1}{2}$ | | $-1\dfrac{1}{4} \cdot \left(-2\dfrac{1}{2}\right)$ | |

    **c.** Do the rules for multiplying integers apply to all rational numbers? Explain
        your reasoning.

**Notetaking with Vocabulary**

**Vocabulary:**

**Notes:**

## 2.4 Self-Assessment

**Use the scale below to rate your understanding of the learning target and the success criteria.**

| *1* | *2* | *3* | *4* |
|---|---|---|---|
| I do not understand. | I can do it with help. | I can do it on my own. | I can teach someone else. |

|  | Rating | Date |
|---|---|---|
| **2.4 Multiplying Rational Numbers** | | |
| **Learning Target:** Find products of rational numbers. | 1  2  3  4 | |
| I can explain the rules for multiplying rational numbers. | 1  2  3  4 | |
| I can find products of rational numbers with the same sign. | 1  2  3  4 | |
| I can find products of rational numbers with different signs. | 1  2  3  4 | |

## 2.4 Practice

**Find the product. Write fractions in simplest form.**

1. $1\frac{2}{3}\left(-2\frac{9}{10}\right)$

2. $-\left(3\frac{2}{5}\right)^2$

3. $-1.27 \cdot (-2.02)$

4. Mixed nuts cost $7.80 per pound. You purchase $2\frac{1}{4}$ pounds. What is the total cost of your purchase?

5. Write two fractions whose product is $-\frac{4}{9}$.

**Evaluate the expression. Write fractions in simplest form.**

6. $-0.2^3 - 4.15(-0.06)$

7. $2\frac{1}{3} \times \left(-4\frac{5}{7}\right) - \left(-\frac{3}{5}\right)^2$

8. A gallon of gasoline costs $2.96. Your car has a 25-gallon gas tank.

   a. Find the cost of filling your gas tank if it is already $\frac{3}{8}$ full.

   b. Your car averages $17\frac{3}{5}$ miles per gallon. After introducing an additive to your gasoline, your car averages $19\frac{1}{5}$ miles per gallon. The company advertised that the additive would give a 10% increase in miles per gallon. Did your car have a 10% increase in miles per gallon? Justify your answer.

9. Use positive or negative integers to fill in the blanks so that the product is $-\frac{1}{6}$. Justify your answer.

$$\frac{\square}{\square} \times \frac{-2}{\square} \times \left(-\frac{\square}{3}\right)$$

10. A school record for the 5K run is 20.15 minutes. Predict the school record after 12 years when the school record decreases by about 0.155 second per year.

11. Use positive or negative integers to fill in the blanks so that the product is –46.25.

$$1\square.5 \times (-3.7\square)$$

# 2.5 Dividing Rational Numbers
**For use with Exploration 2.5**

**Learning Target:** Find quotients of rational numbers.

**Success Criteria:**
- I can explain the rules for dividing rational numbers.
- I can find quotients of rational numbers with the same sign.
- I can find quotients of rational numbers with different signs.

---

**1 EXPLORATION: Finding Quotients of Rational Numbers**

**Work with a partner.**

**a.** Write two division expressions represented by the area model. Then find the quotients.

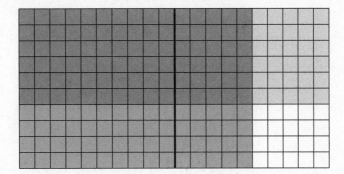

---

**2.5**   **Dividing Rational Numbers** (continued)

**b.** Complete the table.

| | Expression | Quotient | Expression | Quotient |
|---|---|---|---|---|
| **i.** | $0.9 \div 1.5$ | | $-0.9 \div 1.5$ | |
| **ii.** | $1 \div \dfrac{1}{2}$ | | $-1 \div \dfrac{1}{2}$ | |
| **iii.** | $2 \div 0.25$ | | $2 \div (-0.25)$ | |
| **iv.** | $0 \div \dfrac{4}{5}$ | | $0 \div \left(-\dfrac{4}{5}\right)$ | |
| **v.** | $1\dfrac{1}{2} \div 3$ | | $-1\dfrac{1}{2} \div (-3)$ | |
| **vi.** | $0.8 \div 0.1$ | | $-0.8 \div (-0.1)$ | |

**c.** Do the rules for dividing integers apply to all rational numbers? Explain your reasoning.

**d.** Write a real-life story involving the quotient $-0.75 \div 3$. Interpret the quotient in the context of the story.

# 2.5 Notetaking with Vocabulary

**Vocabulary:**

**Notes:**

# 2.5 Self-Assessment

Use the scale below to rate your understanding of the learning target and the success criteria.

| **1** | **2** | **3** | **4** |
|-------|-------|-------|-------|
| I do not understand. | I can do it with help. | I can do it on my own. | I can teach someone else. |

|  | Rating | Date |
|---|---|---|
| **2.5 Dividing Rational Numbers** | | |
| **Learning Target:** Find quotients of rational numbers. | 1  2  3  4 | |
| I can explain the rules for dividing rational numbers. | 1  2  3  4 | |
| I can find quotients of rational numbers with the same sign. | 1  2  3  4 | |
| I can find quotients of rational numbers with different signs. | 1  2  3  4 | |

**Big Ideas Math: Modeling Real Life Grade 7 Accelerated**
Student Journal
**45**

Name _____ Date _____

## 2.5 Practice

**Find the quotient. Write fractions in simplest form.**

1. $1\frac{5}{6} \div (-30)$

2. $-2\frac{4}{5} \div 10\frac{2}{3}$

3. $14.616 \div (-2.32)$

4. How many three-quarter pound burgers can be made with twelve pounds of hamburger?

5. The table shows the changes in your times (in seconds) at the new skateboard ramp.

   | Trial | 1 | 2 | 3 | 4 | 5 |
   |---|---|---|---|---|---|
   | Change | 2.2 | −1.4 | 0.6 | −2.3 | −1.7 |

   a. What is your mean change?

   b. After the 6th trial, your mean change was −0.35. What was the change in your time on the 6th trial?

   c. Was your 6th trial faster or slower than your 5th trial? Explain.

**Evaluate the expression. Write fractions in simplest form.**

6. $5 - 3\frac{9}{10} \div 2\frac{3}{5}$

7. $-3.4^2 \div 1.7 + 7.5$

8. $\dfrac{5\frac{2}{3}}{2\frac{1}{6}}$

9. $\dfrac{-\frac{3}{10}}{1-\frac{2}{5}}$

10. $\dfrac{\frac{15}{8}}{\frac{5}{2}-\frac{1}{4}}$

11. $\dfrac{2\frac{2}{3}+1\frac{5}{6}}{\left(\frac{4}{5}\right)\left(-\frac{9}{2}\right)}$

12. A gallon of gasoline costs $2.96. Your car can travel 28.8 miles on each gallon of gasoline.

   a. You take a trip of length $705\frac{3}{5}$ miles. How much money do you spend on gasoline?

   b. You take the same trip the next year, with the same car. You spend $90.65 on gasoline. What is the percent increase in the price per gallon of gasoline?

# Chapter Self-Assessment

Use the scale below to rate your understanding of the learning target and the success criteria.

**1** I do not understand.

**2** I can do it with help.

**3** I can do it on my own.

**4** I can teach someone else.

| | Rating | Date |
|---|---|---|
| **2.1 Multiplying Integers** | | |
| **Learning Target:** Find products of integers. | 1  2  3  4 | |
| I can explain the rules for multiplying integers. | 1  2  3  4 | |
| I can find products of integers with the same sign. | 1  2  3  4 | |
| I can find products of integers with different signs. | 1  2  3  4 | |
| **2.2 Dividing Integers** | | |
| **Learning Target:** Find quotients of integers. | 1  2  3  4 | |
| I can explain the rules for dividing integers. | 1  2  3  4 | |
| I can find quotients of integers with the same sign. | 1  2  3  4 | |
| I can find quotients of integers with different signs. | 1  2  3  4 | |
| **2.3 Converting Between Fractions and Decimals** | | |
| **Learning Target:** Convert between different forms of rational numbers. | 1  2  3  4 | |
| I can explain the difference between terminating and repeating decimals. | 1  2  3  4 | |
| I can write fractions and mixed numbers as decimals. | 1  2  3  4 | |
| I can write decimals as fractions and mixed numbers. | 1  2  3  4 | |

**Big Ideas Math: Modeling Real Life Grade 7 Accelerated**
Student Journal

**47**

## Chapter 2 — Chapter Self-Assessment (continued)

| | Rating | Date |
|---|---|---|
| **2.4 Multiplying Rational Numbers** | | |
| **Learning Target:** Find products of rational numbers. | 1  2  3  4 | |
| I can explain the rules for multiplying rational numbers. | 1  2  3  4 | |
| I can find products of rational numbers with the same sign. | 1  2  3  4 | |
| I can find products of rational numbers with different signs. | 1  2  3  4 | |
| **2.5 Dividing Rational Numbers** | | |
| **Learning Target:** Find quotients of rational numbers. | 1  2  3  4 | |
| I can explain the rules for dividing rational numbers. | 1  2  3  4 | |
| I can find quotients of rational numbers with the same sign. | 1  2  3  4 | |
| I can find quotients of rational numbers with different signs. | 1  2  3  4 | |

## Chapter 3 Review & Refresh

**Evaluate the expression when $x = \frac{1}{2}$ and $y = -3$.**

**1.** $-4xy$

**2.** $6x - 3y$

**3.** $-5y + 8x + 1$

**4.** $-x^2 - y + 2$

**5.** $2x + 4\left(x + \frac{1}{2}\right) + 7$

**6.** $2(4x - 1)^2 - 3$

**7.** Find the area of the garden when $x = 2$ feet.

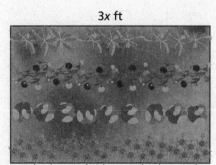

3x ft

2x ft

**Chapter 3** **Review & Refresh** (continued)

**Find the product or quotient. Write fractions in simplest form.**

8. $-\dfrac{3}{10} \times \dfrac{5}{9}$

9. $-\dfrac{4}{11} \times \left(-4\dfrac{1}{8}\right)$

10. $1\dfrac{1}{5} \div -\dfrac{2}{15}$

11. $-2\dfrac{2}{9} \div 2\dfrac{1}{12}$

**Find the sum or difference. Write fractions in simplest form.**

12. $\dfrac{6}{25} - \left(-\dfrac{1}{5}\right)$

13. $-5\dfrac{7}{8} - \left(-2\dfrac{1}{3}\right)$

14. $9.7 - 12.4$

15. $-1.9 + 13.8$

## 3.1 Algebraic Expressions
### For use with Exploration 3.1

**Learning Target:** Simplify algebraic expressions.

**Success Criteria:**
- I can identify terms and like terms of algebraic expressions.
- I can combine like terms to simplify algebraic expressions.
- I can write and simplify algebraic expressions to solve real-life problems.

**1 EXPLORATION: Simplifying Algebraic Expressions**

**Work with a partner.**

**a.** Choose a value of $x$ other than 0 or 1 for the last column in the table. Complete the table by evaluating each algebraic expression for each value of $x$. What do you notice?

| | Expression | Value When $x = 0$ | $x = 1$ | $x = ?$ |
|---|---|---|---|---|
| **A.** | $-\frac{1}{3} + x + \frac{7}{3}$ | | | |
| **B.** | $0.5x + 3 - 1.5x - 1$ | | | |
| **C.** | $2x + 6$ | | | |
| **D.** | $x + 4$ | | | |
| **E.** | $-2x + 2$ | | | |
| **F.** | $\frac{1}{2}x - x + \frac{3}{2}x + 4$ | | | |
| **G.** | $-4.8x + 2 - x + 3.8x$ | | | |
| **H.** | $x + 2$ | | | |
| **I.** | $-x + 2$ | | | |
| **J.** | $3x + 2 - x + 4$ | | | |

**3.1** **Algebraic Expressions** (continued)

**b.** How can you use properties of operations to justify your answers in part (a)? Explain your reasoning.

**c.** To subtract a number, you can add its opposite. Does a similar rule apply to the terms of an algebraic expression? Explain your reasoning.

## 3.1 Notetaking with Vocabulary

**Vocabulary:**

**Notes:**

## 3.1 Self-Assessment

**Use the scale below to rate your understanding of the learning target and the success criteria.**

| 1 | 2 | 3 | 4 |
|---|---|---|---|
| I do not understand. | I can do it with help. | I can do it on my own. | I can teach someone else. |

| | Rating | Date |
|---|---|---|
| **3.1 Algebraic Expressions** | | |
| **Learning Target:** Simplify algebraic expressions. | 1  2  3  4 | |
| I can identify terms and like terms of algebraic expressions. | 1  2  3  4 | |
| I can combine like terms to simplify algebraic expressions. | 1  2  3  4 | |
| I can write and simplify algebraic expressions to solve real-life problems. | 1  2  3  4 | |

Name _____ Date _____

## 3.1 Practice

**Identify the terms and like terms in the expression.**

**1.** $1.3x - 2.7x^2 - 5.4x + 3$

**2.** $10 - \frac{3}{10}m + 6m^2 + \frac{2}{5}m$

**Simplify the expression.**

**3.** $-\frac{15}{4}b + \frac{5}{6}b$

**4.** $9y - 15y + 12 - 6y$

**5.** Write an expression in simplest form that represents the perimeter of the polygon.

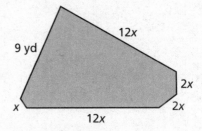

9 yd

12x

2x

2x

x

12x

**6.** Danielle is $x$ years old. Her sister is 5 years older and her brother is half Danielle's age. Write an expression in simplest form for the sum of their ages.

**7.** You buy $x$ packs of pencils, twice as many packs of erasers, and three times as many rolls of tape. Write an expression in simplest form for the total amount of money you spent.

**8.** Evaluate the expression
$-\frac{2}{3}x - 4 + \frac{1}{2}x - 3 + \frac{5}{6}x$ when $x = 3$
before and after simplifying. Which method do you prefer, *before simplifying* or *after simplifying*? Explain.

**9.** Write an expression with seven different terms that is equivalent to $9x - 4x^2 - 15 + y$. Justify your answer.

**10.** For what value(s) of $x$ is the expression $3x - \frac{1}{4} + 5$ equivalent to the expression $-\frac{1}{4} + 5$? Explain.

## 3.2 Adding and Subtracting Linear Expressions
**For use with Exploration 3.2**

**Learning Target:** Find sums and differences of linear expressions.

**Success Criteria:**
- I can explain the difference between linear and nonlinear expressions.
- I can find opposites of terms that include variables.
- I can apply properties of operations to add and subtract linear expressions.

**1  EXPLORATION:** Using Algebra Tiles

**Work with a partner. You can use algebra tiles shown at the right to find sums and differences of algebraic expressions.**

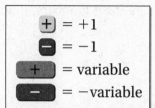

a. How can you use algebra tiles to model a sum of terms that equals 0? Explain your reasoning.

b. Write each sum or difference modeled below. Then use the algebra tiles to simplify the expression.

$$\left( \boxed{+} \; \boxed{+}\boxed{+}\boxed{+}\boxed{+} \right) + \left( \boxed{+} \atop \boxed{+} \; \boxed{-}\boxed{-}\boxed{-} \right)$$

$$\left( \boxed{+} \atop \boxed{+} \; \boxed{-}\boxed{-}\boxed{-}\boxed{-} \atop \boxed{-}\boxed{-}\boxed{-} \right) + \left( \boxed{-} \atop \boxed{-} \atop \boxed{-} \; \boxed{+}\boxed{+}\boxed{+} \atop \boxed{+}\boxed{+} \right)$$

$$\left( \boxed{+} \; \boxed{-}\boxed{-}\boxed{-}\boxed{-} \right) - \left( \boxed{+} \; \boxed{-}\boxed{-}\boxed{-} \right)$$

**3.2**  **Adding and Subtracting Linear Expressions** (continued)

$$\left( \boxed{\begin{array}{c}-\\-\end{array}} \ \boxed{+}\boxed{+}\boxed{+}\boxed{+}\boxed{+} \right) - \left( \boxed{+}\boxed{-} \right)$$

**c.** Write two algebraic expressions of the form $ax + b$, where $a$ and $b$ are rational numbers. Find the sum and difference of the expressions.

**2**  **EXPLORATION:** Using Properties of Operations

**Work with a partner.**

**a.** Do algebraic expressions, such as $2x$, $-3y$, and $3z + 1$, have additive inverses? How do you know?

**b.** How can you find the sums and differences modeled in Exploration 1 without using algebra tiles? Explain your reasoning.

## 3.2 Notetaking with Vocabulary

**Vocabulary:**

**Notes:**

## 3.2 Self-Assessment

Use the scale below to rate your understanding of the learning target and the success criteria.

| 1 | 2 | 3 | 4 |
|---|---|---|---|
| I do not understand. | I can do it with help. | I can do it on my own. | I can teach someone else. |

|  | Rating | Date |
|---|---|---|
| **3.2 Adding and Subtracting Linear Expressions** | | |
| **Learning Target:** Find sums and differences of linear expressions. | 1   2   3   4 | |
| I can explain the difference between linear and nonlinear expressions. | 1   2   3   4 | |
| I can find opposites of terms that include variables. | 1   2   3   4 | |
| I can apply properties of operations to add and subtract linear expressions. | 1   2   3   4 | |

Name _____ Date _____

## 3.2 Practice

**Find the sum.**

1. $(-1.2t + 5) + (3t + 2)$

2. $\left(\frac{7}{4}k + 9\right) + \left(\frac{1}{3}k - 3\right)$

3. $(-0.3s - 2) + (5 - 3.4s) + (5r + 2)$

4. $\left(\frac{1}{6}p - 3\right) + (3q - 1.7) + \left(-\frac{1}{3}p + 4\right)$

5. You are collecting pairs of socks and toothbrushes for a local charity. After $d$ days, you have collected $(4d + 5)$ pairs of socks and $(3d + 7)$ toothbrushes.

   a. Write an expression that represents the total number of items that have been collected.

   b. How many more pairs of socks than toothbrushes have been collected on day 7?

**Find the difference.**

6. $(8.2k - 6) - (4.5 - 9k)$

7. $\left(-\frac{3}{5} + 2g\right) - \left(4 + \frac{2}{3}g\right)$

8. $\left(\frac{2}{7}x - 3\right) - \left(\frac{5}{2} + \frac{1}{5}y\right)$

9. $(3 - 2.1z) - (1.7w + 9) - (-8 - 3z)$

10. The expression $7n - 19$ represents the perimeter of the triangle.

    a. What is the length of the third side?

    b. What are the lengths of the three sides of the triangle when $n = 5$?

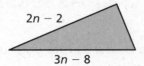
$2n - 2$
$3n - 8$

    c. What are the lengths of the three sides of the triangle when $n = 7$?

11. Consider $ax + b$ and $cx + d$, two linear expressions.

    a. Find values for $a$, $b$, $c$, and $d$, such that the expression $(ax + b) - (cx + d)$ is a linear expression.

    b. Find values for $a$, $b$, $c$, and $d$, such that the expression $(ax + b) - (cx + d)$ is not a linear expression.

Name_____ Date_____

# 3.3 The Distributive Property
### For use with Exploration 3.3

**Learning Target:** Apply the Distributive Property to generate equivalent expressions.

**Success Criteria:** • I can explain how to apply the Distributive Property.
• I can use the Distributive Property to simplify algebraic expressions.

---

**1 EXPLORATION:** Using Models to Write Expressions

**Work with a partner.**

a. Write an expression that represents the area of the shaded region in each figure.

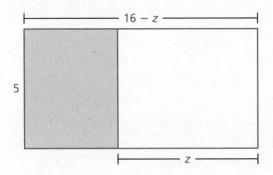

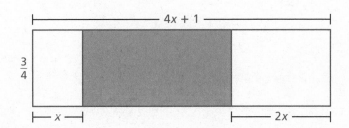

**3.3** **The Distributive Property** (continued)

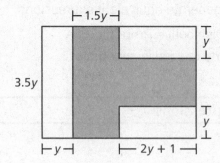

**b.** Compare your expressions in part (a) with other groups in your class. Did other groups write expressions that look different than yours? If so, determine whether the expressions are equivalent.

**Notetaking with Vocabulary**

**Vocabulary:**

**Notes:**

## 3.3 Self-Assessment

Use the scale below to rate your understanding of the learning target and the success criteria.

| 1 | 2 | 3 | 4 |
|---|---|---|---|
| I do not understand. | I can do it with help. | I can do it on my own. | I can teach someone else. |

| | Rating | Date |
|---|---|---|
| **3.3 The Distributive Property** | | |
| **Learning Target:** Apply the Distributive Property to generate equivalent expressions. | 1  2  3  4 | |
| I can explain how to apply the Distributive Property. | 1  2  3  4 | |
| I can use the Distributive Property to simplify algebraic expressions. | 1  2  3  4 | |

## 3.3 Practice

**Simplify the expression.**

1. $60m - 15(4 - 8m) + 20$

2. $4(5.8 - 9x) + 8.2 + 22x$

3. $\frac{5}{3}(5x + 9) + \frac{4}{5}(1 - 9x)$

4. $4(2.5g - 4) + 3(1.2g - 2)$

5. $\frac{1}{5}(-15w - 20) + \frac{1}{2}(3 - 4w)$

6. $\frac{1}{8}(16k - 24) + \frac{1}{5}(2 + 10k)$

7. $(12 + 7.2b) - 3(0.9b - 4)$

8. $\frac{1}{4}(16j - 12) - \frac{1}{9}(18j + 45)$

9. You are selling tickets to a play. You have sold $(3t + 2)$ tickets for $5 each and $(2t + 5)$ tickets for $7 each.

   a. Write an expression that represents the total number of tickets sold so far. Then simplify the expression.

   b. Write an expression that represents the total amount of money received for the tickets that have been sold. Then simplify the expression.

   c. When $t = 3$, what is the total amount of money received?

**Draw a diagram that shows how the expression can represent the area of a figure. Then simplify the expression.**

10. $8(3x - 1)$

11. $(5 + 2)(x + 3x)$

12. The length of a rectangular field is 30 more than twice its width. Write an expression in simplest form for the perimeter of the field in terms of its width $w$.

13. Which of the following is not equivalent? Explain.

   **A.** $5(2x - 7)$       **B.** $(2x - 7)(5)$       **C.** $5(7 - 2x)$       **D.** $5(-7 + 2x)$

14. Your friend asks you to perform the following steps.

   1) Pick any number except 0.

   2) Subtract 5 from your number.

   3) Multiply the result by 2.

   4) Add 10 to the result.

   5) Divide the result by 4. Tell me your result.

   Your friend says, "Your original number was ____!" Explain how your friend knew your original number.

## 3.4 Factoring Expressions
**For use with Exploration 3.4**

**Learning Target:** Factor algebraic expressions.

**Success Criteria:**
- I can identify the greatest common factor of terms, including variable terms.
- I can use the Distributive Property to factor algebraic expressions.
- I can write a term as a product involving a given factor.

---

**1** **EXPLORATION:** Finding Dimensions

**Work with a partner.**

a. The models show the areas (in square units) of parts of rectangles. Use the models to find the missing values that complete the expressions. Explain your reasoning.

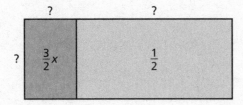

$$\frac{4}{5} + \frac{8}{5} = ?\,(? + ?)$$

$$\frac{3}{2}x + \frac{1}{2} = ?\,(? + ?)$$

### 3.4 Factoring Expressions (continued)

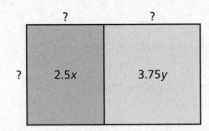

$$2.5x + 3.75y = ?(? + ?)$$

**b.** Are the expressions you wrote in part (a) equivalent to the original expressions? Explain your reasoning.

**c.** Explain how you can use the Distributive Property to find rational number factors of an expression.

## 3.4 Notetaking with Vocabulary

**Vocabulary:**

**Notes:**

## 3.4 Self-Assessment

**Use the scale below to rate your understanding of the learning target and the success criteria.**

| 1 | 2 | 3 | 4 |
|---|---|---|---|
| I do not understand. | I can do it with help. | I can do it on my own. | I can teach someone else. |

|  | Rating | Date |
|---|---|---|
| **3.4 Factoring Expressions** | | |
| **Learning Target:** Factor algebraic expressions. | 1  2  3  4 | |
| I can identify the greatest common factor of terms, including variable terms. | 1  2  3  4 | |
| I can use the Distributive Property to factor algebraic expressions. | 1  2  3  4 | |
| I can write a term as a product involving a given factor. | 1  2  3  4 | |

## 3.4 Practice

**Factor the expression using the GCF.**

1. $21s + 15$

2. $32v + 24w$

3. $12y - 42z$

**Factor out the coefficient of the variable term.**

4. $1.2k + 2.4$

5. $3f + 5$

6. $\frac{3}{10}x - \frac{3}{5}$

7. Factor $-\frac{1}{3}$ out of $-\frac{1}{3}x - 12$.

8. Factor $-\frac{1}{6}$ out of $-\frac{1}{3}x + \frac{5}{6}y$.

9. The area of the rectangle is $(18x - 12)$ square inches. Write an expression that represents the length of the rectangle (in inches).

6 in.

10. A concession stand sells hamburgers. The revenue from the hamburgers is $(30x + 45)$ dollars.

   a. The price of a hamburger is \$5. Write an expression that represents the number of hamburgers sold.

   b. The revenue from drinks is $(63x + 84)$ dollars. The price of a drink is \$3. Write an expression that represents the number of drinks sold.

   c. Write and simplify an expression that represents how many more drinks than hamburgers were sold.

11. The formula for the area of a triangle is $A = \frac{1}{2}bh$. A triangle has an area of $\left(\frac{1}{12}x + \frac{7}{30}\right)$.

   a. Write a multiplication expression that can represent the product of $\frac{1}{2}$ and the base and the height of the triangle.

   b. Write a multiplication expression that can represent the product of $\frac{1}{2}$ and the base and the height of the triangle, where the factor containing the x-term does not have fractions.

Name_____ Date_____

## Chapter 3 Chapter Self-Assessment

Use the scale below to rate your understanding of the learning target and the success criteria.

| 1 | 2 | 3 | 4 |
|---|---|---|---|
| I do not understand. | I can do it with help. | I can do it on my own. | I can teach someone else. |

|  | Rating | Date |
|---|---|---|
| **3.1 Algebraic Expressions** | | |
| **Learning Target:** Simplify algebraic expressions. | 1  2  3  4 | |
| I can identify terms and like terms of algebraic expressions. | 1  2  3  4 | |
| I can combine like terms to simplify algebraic expressions. | 1  2  3  4 | |
| I can write and simplify algebraic expressions to solve real-life problems. | 1  2  3  4 | |
| **3.2 Adding and Subtraction Linear Expressions** | | |
| **Learning Target:** Find sums and differences of linear expressions. | 1  2  3  4 | |
| I can explain the difference between linear and nonlinear expressions. | 1  2  3  4 | |
| I can find opposites of terms that include variables. | 1  2  3  4 | |
| I can apply properties of operations to add and subtract linear expressions. | 1  2  3  4 | |
| **3.3 The Distributive Property** | | |
| **Learning Target:** Apply the Distributive Property to generate equivalent expressions. | 1  2  3  4 | |
| I can explain how to apply the Distributive Property. | 1  2  3  4 | |
| I can use the Distributive Property to simplify algebraic expressions. | 1  2  3  4 | |

# Chapter Self-Assessment (continued)

| | Rating | Date |
|---|---|---|
| **3.4 Factoring Expressions** | | |
| **Learning Target:** Factor algebraic expressions. | 1  2  3  4 | |
| I can identify the greatest common factor of terms, including variable terms. | 1  2  3  4 | |
| I can use the Distributive Property to factor algebraic expressions. | 1  2  3  4 | |
| I can write a term as a product involving a given factor. | 1  2  3  4 | |

Name_____ Date _____

**Write the phrase as an algebraic expression.**

1. the sum of eight and a number $y$

2. six less than a number $p$

3. the product of seven and a number $m$

4. eight less than the product of eleven and a number $c$

5. a number $r$ decreased by the quotient of a number $r$ and two

6. the product of nine and the sum of a number $z$ and four

**Big Ideas Math: Modeling Real Life Grade 7 Accelerated** **69**

**Chapter 4** **Review & Refresh** (continued)

**Complete the number sentence with < or >.**

**7.** $\frac{3}{4}$ ___ 0.2

**8.** $\frac{7}{8}$ ___ 0.7

**9.** $-0.6$ ___ $-\frac{2}{3}$

**10.** $-1.76$ ___ 1.75

**11.** $\frac{17}{3}$ ___ 6

**12.** 1.8 ___ $\frac{31}{16}$

**13.** Your height is 5 feet and $1\frac{5}{8}$ inches. Your friend's height is 5.6 feet. Who is taller? Explain.

# 4.1 Solving Equations Using Addition or Subtraction
**For use with Exploration 4.1**

**Learning Target:** Write and solve equations using addition or subtraction.

**Success Criteria:**
- I can apply the Addition and Subtraction Properties of Equality to produce equivalent equations.
- I can solve equations using addition or subtraction.
- I can apply equations involving addition or subtraction to solve real-life problems.

---

**1 EXPLORATION:** Using Algebra Tiles to Solve Equations

**Work with a partner.**

a. Use the examples to explain the meaning of each property.

**Addition Property of Equality:**
$$x + 2 = 1$$
$$x + 2 + 5 = 1 + 5$$

**Subtraction Property of Equality:**
$$x + 2 = 1$$
$$x + 2 - 1 = 1 - 1$$

Are these properties true for equations involving negative numbers? Explain your reasoning.

**4.1** Solving Equations Using Addition or Subtraction (continued)

**b.** Write the four equations modeled by the algebra tiles. Explain how you can use algebra tiles to solve each equation. Then find the solutions.

**c.** How can you solve each equation in part (b) without using algebra tiles?

#  4.1 Notetaking with Vocabulary

**Vocabulary:**

**Notes:**

# 4.1 Self-Assessment

Use the scale below to rate your understanding of the learning target and the success criteria.

| **1** | **2** | **3** | **4** |
|---|---|---|---|
| I do not understand. | I can do it with help. | I can do it on my own. | I can teach someone else. |

| | Rating | Date |
|---|---|---|
| **4.1 Solving Equations Using Addition or Subtraction** | | |
| **Learning Target:** Write and solve equations using addition or subtraction. | 1  2  3  4 | |
| I can apply the Addition and Subtraction Properties of Equality to produce equivalent equations. | 1  2  3  4 | |
| I can solve equations using addition or subtraction. | 1  2  3  4 | |
| I can apply equations involving addition or subtraction to solve real-life problems. | 1  2  3  4 | |

## 4.1 Practice

**Solve the equation. Check your solution.**

1. $-4.5 = m + 1.9$

2. $10\frac{2}{3} = r + 12\frac{1}{6}$

3. $b + 4.006 = 9$

4. $-7\frac{5}{8} = 1\frac{5}{6} + d$

5. $f - \frac{2}{15} = 6\frac{3}{5}$

6. $-10.216 + c = -12.014$

**Write the word sentence as an equation. Then solve the equation.**

7. 27 is 12 more than a number $x$.

8. The difference of a number $p$ and $-9$ is 12.

9. 35 less than a number $m$ is $-72$.

10. You swim the 50-meter freestyle in 28.12 seconds. This is 0.14 second less than your previous fastest time. Write and solve an equation to find your previous fastest time.

11. The perimeter of a rectangular backyard is $32\frac{1}{2}$ meters. The two shorter sides are each $7\frac{3}{8}$ meters long. Write and solve an equation to find the length of the two longer sides? (*Hint:* The sum of the shorter side and the longer side is equal to half of the perimeter.)

12. The wind chill when the airplane lands in Minneapolis is $-70.2°F$, which is $140.8°F$ less than the wind chill when the airplane took off in Orlando. Write and solve an equation to find the wind chill when the airplane took off in Orlando.

13. Your cell phone bill in August was $61.43, which was $21.75 more than your bill in July. Your cell phone bill in July was $13.62 less than your bill in June. What was your cell phone bill in June?

**Solve the equation.**

14. $|x| - 10.5 = 4.3$

15. $|x + 2| - 7 = 5$

16. Find the value of $3x + 2$ when $x - 5 = -8$.

17. Find the values of $x - 4$ when $|x| + 1 = 3$.

Name_____ Date _____

## 4.2 Solving Equations Using Multiplication or Division
**For use with Exploration 4.2**

**Learning Target:** Write and solve equations using multiplication or division.

**Success Criteria:**
- I can apply the Multiplication and Division Properties of Equality to produce equivalent equations.
- I can solve equations using multiplication or division.
- I can apply equations involving multiplication or division to solve real-life problems.

**1  EXPLORATION:** Using Algebra Tiles to Solve Equations

**Work with a partner.**

a. Use the examples to explain the meaning of each property.

**Multiplication Property of Equality:** $3x = 1$
$$2 \cdot 3x = 2 \cdot 1$$

**Division Property of Equality:** $3x = 1$
$$\frac{3x}{4} = \frac{1}{4}$$

Are these properties true for equations involving negative numbers? Explain your reasoning.

**4.2** **Solving Equations Using Multiplication or Division** (continued)

**b.** Write the three equations modeled by the algebra tiles. Explain how you can use algebra tiles to solve each equation. Then find the solutions.

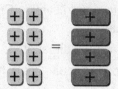

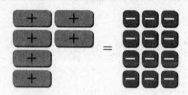

**c.** How can you solve each equation in part (b) without using algebra tiles?

## 4.2 Notetaking with Vocabulary

**Vocabulary:**

**Notes:**

## 4.2 Self-Assessment

Use the scale below to rate your understanding of the learning target and the success criteria.

| 1 | 2 | 3 | 4 |
|---|---|---|---|
| I do not understand. | I can do it with help. | I can do it on my own. | I can teach someone else. |

| | Rating | Date |
|---|---|---|
| **4.2 Solving Equations Using Multiplication or Division** | | |
| **Learning Target:** Write and solve equations using multiplication or division. | 1   2   3   4 | |
| I can apply the Multiplication and Division Properties of Equality to produce equivalent equations. | 1   2   3   4 | |
| I can solve equations using multiplication or division. | 1   2   3   4 | |
| I can apply equations involving multiplication or division to solve real-life problems. | 1   2   3   4 | |

## 4.2 Practice

**Solve the equation. Check your solution.**

1. $-14p = -21$

2. $-\dfrac{8}{15}k = -4$

3. $-7.24q = 17.014$

4. $\dfrac{g}{0.003} = -2.8$

5. $-\dfrac{10}{21}c = -\dfrac{15}{28}$

6. $18 = -\dfrac{6}{11}h$

7. You order an entrée for $12.00. You pay $0.78 in taxes. Write and solve an equation to determine the tax rate.

8. If a project is handed in late, you receive $\dfrac{8}{9}$ of your earned points. You received 72 points on your late project. Write and solve an equation to determine how many points you lost.

9. Write a multiplication equation that has a solution of $-14.8$.

10. Write a division equation that has a solution of $-\dfrac{9}{14}$.

**Write the word sentence as an equation. Then solve the equation.**

11. A number divided by $\dfrac{1}{2}$ is 10.

12. The product of a number and $-12$ is $-8$.

13. The quotient of 9.75 and a number is $-6.5$

14. There are 92 students in a room. They are separated into 18 groups. How many students are in each group? How many students are not in a group?

15. A bus token costs $1.75.

    a. You spend $15.75 on tokens. Write and solve an equation to find how many tokens you purchase.

    b. If you purchase 10 tokens, you get 2 free tokens. Write and solve an equation to find the approximate reduced price of each token when purchasing this special.

    c. You also receive free tokens if you purchase 20 tokens. The reduced price for each token is $1.40. Write and solve an equation to find how many free tokens you receive when purchasing this special.

16. Solve $\dfrac{1}{3}|z| = 2$.

## 4.3 Solving Two-Step Equations
**For use with Exploration 4.3**

**Learning Target:** Write and solve two-step equations.

**Success Criteria:**
- I can apply properties of equality to produce equivalent equations.
- I can solve two-step equations using the basic operations.
- I can apply two-step equations to solve real-life problems.

---

**1 EXPLORATION:** Using Algebra Tiles to Solve Equations

**Work with a partner.**

a. What is being modeled by the algebra tiles below? What is the solution?

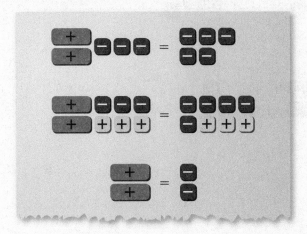

b. Use properties of equality to solve the original equation in part (a). How do your steps compare to the steps performed with algebra tiles?

**Big Ideas Math: Modeling Real Life Grade 7 Accelerated**
Student Journal

**79**

**4.3** **Solving Two-Step Equations** (continued)

**c.** Write the three equations modeled by the algebra tiles below. Then solve each equation using algebra tiles. Check your answers using properties of equality.

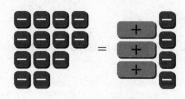

**d.** Explain how to solve an equation of the form $ax + b = c$ for $x$.

**Notetaking with Vocabulary**

**Vocabulary:**

**Notes:**

## 4.3 Self-Assessment

Use the scale below to rate your understanding of the learning target and the success criteria.

| **1** | **2** | **3** | **4** |
|---|---|---|---|
| I do not understand. | I can do it with help. | I can do it on my own. | I can teach someone else. |

|  | Rating | Date |
|---|---|---|
| **4.3 Solving Two-Step Equations** | | |
| **Learning Target:** Write and solve two-step equations. | 1  2  3  4 | |
| I can apply properties of equality to produce equivalent equations. | 1  2  3  4 | |
| I can solve two-step equations using the basic operations. | 1  2  3  4 | |
| I can apply two-step equations to solve real-life problems. | 1  2  3  4 | |

Name _____ Date _____

## 4.3 Practice

**Solve the equation. Check your solution.**

1. $25 = 4.5z + 12$
2. $5.25s - 2.01 = -8.94$
3. $16 + 2.4c = 22.5$

4. $4h + \dfrac{1}{3} = \dfrac{3}{4}$
5. $\dfrac{1}{7}f - 5\dfrac{1}{2} = \dfrac{9}{14}$
6. $-\dfrac{1}{2}u + \dfrac{3}{5} = \dfrac{1}{6}$

7. You purchased $132.49 worth of wheels and bearings for your skateboards. The shop charges $15 per board to install them. The total cost is $192.49. Write and solve an equation to determine how many skateboards will be repaired.

8. A music download service charges a flat fee each month and $0.99 per download. The total cost for downloading 27 songs this month is $42.72. Write and solve an equation to determine the flat fee.

**Solve the equation. Check your solution.**

9. $-5(m + 4) = 27$
10. $-12(a - 2) = -50$
11. $-5x - 2x + 3x = 9$

12. The perimeter of a triangle is 60 feet. One leg is 12 feet long. Of the two unknown sides, one the them is twice as long as the other. Find the lengths of the two unknown sides.

13. Sally picks seashells by the seashore. She lost 17 of them on her way home. With the remaining seashells, she planned to fill 5 jars with the same number of seashells in each. How many seashells did Sally originally pick?

    a. You do not have enough information to solve this problem. The number of seashells in each jar is the same as the number portion of her street address, which is a 2-digit number. The first digit is 5. The last digit is 9 less than 3 times the first digit. How many seashells did Sally plan to put in each jar?

    b. By working backwards, determine how many seashells Sally originally picked.

    c. The 5 jars that Sally chose would not each hold that many seashells. In her search for a 6th jar, she discovered a few seashells in her pocket. What are possible values for the number of seashells in each of the 6 jars and the number of seashells discovered in her pocket, such that there are no seashells left over?

## 4.4 Writing and Graphing Inequalities
**For use with Exploration 4.4**

**Learning Target:** Write inequalities and represent solutions of inequalities on number lines.

**Success Criteria:**
- I can write word sentences as inequalities.
- I can determine whether a value is a solution of an inequality.
- I can graph the solutions of inequalities.

---

**1 EXPLORATION: Understanding Inequality Statements**

**Work with a partner. Create a number line on the floor with both positive and negative numbers.**

a. For each statement, stand at a number on your number line that could represent the situation. On what other numbers can you stand?

At least 3 students from our school are in a chess tournament.

Your ring size is less than 7.5.

The temperature is no more than –1 degree Fahrenheit.

---

**Big Ideas Math: Modeling Real Life Grade 7 Accelerated**

**4.4** **Writing and Graphing Inequalities** (continued)

The elevation of a frogfish is greater than $-8\frac{1}{2}$ meters.

**b.** How can you represent all of the solutions for each statement in part (a) on a number line?

## 4.4 Notetaking with Vocabulary

**Vocabulary:**

**Notes:**

## 4.4 Self-Assessment

Use the scale below to rate your understanding of the learning target and the success criteria.

| *1* | *2* | *3* | *4* |
|---|---|---|---|
| I do not understand. | I can do it with help. | I can do it on my own. | I can teach someone else. |

|  | Rating | Date |
|---|---|---|
| **4.4 Writing and Graphing Inequalities** | | |
| **Learning Target:** Write inequalities and represent solutions of inequalities on number lines. | 1   2   3   4 | |
| I can write word sentences as inequalities. | 1   2   3   4 | |
| I can determine whether a value is a solution of an inequality. | 1   2   3   4 | |
| I can graph the solutions of inequalities. | 1   2   3   4 | |

## 4.4 Practice

**Write an inequality for the graph. Then, in words, describe all the values of *x* that make the inequality true.**

1.

2.

**Write the word sentence as an inequality.**

3. A number *x* is at least 15.

4. A number *r* added to 3.7 is less than 1.2.

5. A number *h* divided by 2 is more than –5.

6. A number *a* minus 8.2 is no greater than 12.

**Tell whether the given value is a solution of the inequality.**

7. $p + 1.7 \geq -4;\ p = -9$

8. $\frac{3}{4} - d > \frac{1}{3};\ d = \frac{1}{2}$

9. $5t < 4 - t;\ t = -3$

10. $\frac{q}{5} < q - 20;\ q = 15$

**Graph the inequality on a number line.**

11. $\ell \leq 3.5$

12. $m > -15$

13. To get a job at the local restaurant, you must be at least 16 years old. Write an inequality that represents this situation.

14. In order to qualify for a college scholarship, you must have acceptable scores in either the SAT or the ACT along with the following requirements: a minimum GPA of 3.5; at least 12 credits of college preparatory academic courses; and at least 75 hours of community service.

    a. Write and graph three inequalities that represent the requirements.

    b. Your cousin has a GPA of 3.6, 15 credits of college preparatory class, and 65 hours of community service. Other than the test scores, does your cousin satisfy the requirements? Explain.

15. Which of the following is not equivalent? Explain.

    A number *x* is at most 3.　　　　　A number *x* is not greater than 3.

    A number *x* is less than 3.　　　　A number *x* is less than or equal to 3.

## 4.5 Solving Inequalities Using Addition or Subtraction
**For use with Exploration 4.5**

**Learning Target:** Write and solve inequalities using addition or subtraction.

**Success Criteria:**
- I can apply the Addition and Subtraction Properties of Inequality to produce equivalent inequalities.
- I can solve inequalities using addition or subtraction.
- I can apply inequalities involving addition or subtraction to solve real-life problems.

### 1 EXPLORATION: Writing Inequalities

**Work with a partner. Use two number cubes on which the odd numbers are negative on one of the number cubes and the even numbers are negative on the other number cube.**

Roll the number cubes. Write an inequality that compares the numbers.

Roll one of the number cubes. Add the number to each side of the inequality and record your result.

Repeat the previous two steps five more times.

**4.5**   **Solving Inequalities Using Addition or Subtraction** (continued)

**a.** When you add the same number to each side of an inequality, does the inequality remain true? Explain your reasoning.

**b.** When you subtract the same number from each side of an inequality, does the inequality remain true? Use inequalities generated by number cubes to justify your answer.

**c.** Use your results in parts (a) and (b) to make a conjecture about how to solve an inequality of the form $x + a < b$ for $x$.

## 4.5 Notetaking with Vocabulary

**Vocabulary:**

**Notes:**

## 4.5 Self-Assessment

Use the scale below to rate your understanding of the learning target and the success criteria.

| 1 | 2 | 3 | 4 |
|---|---|---|---|
| I do not understand. | I can do it with help. | I can do it on my own. | I can teach someone else. |

|  | Rating | Date |
|---|---|---|
| **4.5 Solving Inequalities Using Addition or Subtraction** | | |
| **Learning Target:** Write and solve inequalities using addition or subtraction. | 1   2   3   4 | |
| I can apply the Addition and Subtraction Properties of Inequality to produce equivalent inequalities. | 1   2   3   4 | |
| I can solve inequalities using addition or subtraction. | 1   2   3   4 | |
| I can apply inequalities involving addition or subtraction to solve real-life problems. | 1   2   3   4 | |

Name _____ Date _____

## 4.5 Practice

**Solve the inequality. Graph the solution.**

1. $w - 1.8 < 2.5$

2. $v + \frac{1}{3} > 8$

3. $\frac{2}{5} < \frac{4}{5} + k$

4. $q + \frac{3}{4} \geq -\frac{1}{4}$

5. $7.4 > c + 3.9$

6. $p - 10.2 > 3.5$

7. You and two friends are diving for lobster. The maximum number of lobsters you may have on your boat is 18. You currently have 7 lobsters.

   a. Write and solve an inequality that represents the additional lobsters that you may catch.

   b. Another friend comes on your boat and he has 3 lobsters. You may now have 24 lobsters on your boat. Write and solve an inequality that represents the additional lobsters that you may catch.

   c. How many lobsters is each person allowed to catch?

**Write and solve an inequality that represents x.**

8. The length is greater than the width.

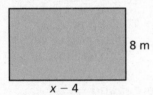

8 m

$x - 4$

9. The perimeter is less than or equal to 50 inches.

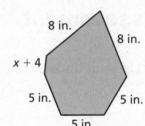

8 in.

8 in.

$x + 4$

5 in.

5 in.

5 in.

10. The solution of $w - c > -3.4$ is $w > -1.4$. What is the value of $c$?

11. Describe all numbers that are solutions to $|x| < 5$.

12. The *triangle inequality theorem* states that the sum of the lengths of any two sides of a triangle is greater than the length of the third side. A triangle has side lengths of 6 inches and 17 inches. What are the possible values for the length of the third side? Explain how you found your answer.

13. The possible values of $x$ are given by $x - 5 \geq 7$. What is the least possible value of $9x$? Explain your reasoning.

# 4.6 Solving Inequalities Using Multiplication or Division
**For use with Exploration 4.6**

**Learning Target:** Write and solve inequalities using multiplication or division.

**Success Criteria:**
- I can apply the Multiplication and Division Properties of Inequality to produce equivalent inequalities.
- I can solve inequalities using multiplication or division.
- I can apply inequalities involving multiplication or division to solve real-life problems.

### 1 EXPLORATION: Writing Inequalities

**Work with a partner. Use two number cubes on which the odd numbers are negative on one of the number cubes and the even numbers are negative on the other number cube.**

Roll the number cubes. Write an inequality that compares the numbers.

Roll one of the number cubes. Multiply each side of the inequality by the number and record your result.

Repeat the previous two steps nine more times.

**4.6** **Solving Inequalities Using Multiplication or Division** (continued)

**a.** When you multiply each side of an inequality by the same number, does the inequality remain true? Explain your reasoning?

**b.** When you divide each side of an inequality by the same number, does the inequality remain true? Use inequalities generated by number cubes to justify your answer.

**c.** Use your results in parts (a) and (b) to make a conjecture about how to solve an inequality of the form $ax < b$ for $x$ when $a > 0$ and when $a < 0$.

## 4.6 Notetaking with Vocabulary

**Vocabulary:**

**Notes:**

## 4.6 Self-Assessment

**Use the scale below to rate your understanding of the learning target and the success criteria.**

| **1** | **2** | **3** | **4** |
|---|---|---|---|
| I do not understand. | I can do it with help. | I can do it on my own. | I can teach someone else. |

| | Rating | Date |
|---|---|---|
| **4.6 Solving Inequalities Using Multiplication or Division** | | |
| **Learning Target:** Write and solve inequalities using multiplication or division. | 1  2  3  4 | |
| I can apply the Multiplication and Division Properties of Inequality to produce equivalent inequalities. | 1  2  3  4 | |
| I can solve inequalities using multiplication or division. | 1  2  3  4 | |
| I can apply inequalities involving multiplication or division to solve real-life problems. | 1  2  3  4 | |

## 4.6 Practice

**Solve the inequality. Graph the solution.**

1. $3y \leq \dfrac{3}{4}$

2. $\dfrac{s}{3.1} \geq 4.5$

3. $-\dfrac{4}{5} < 2x$

4. $-\dfrac{d}{2} > \dfrac{3}{8}$

5. $-1.2 \leq -0.8r$

6. $\dfrac{j}{-5.2} \leq -1.5$

**Write the word sentence as an inequality. Then solve the inequality.**

7. A number divided by 5 is at least 4.

8. The product of 2 and a number is at most –6.

9. The solution of $cx \geq -4$ is $x \geq -8$. What is the value of $c$?

10. The height of a room is 10 feet. You are building shelving from the floor to the ceiling.

   a. Each shelf requires 8 inches. Write and solve an inequality that represents the number of shelves that can be made.

   b. You forgot to include the thickness of each shelf in your measurements. The amount of space needed for each shelf is actually 10 inches. Write and solve an inequality that represents the number of shelves that can be made.

11. Students in a math class are divided into 8 equal groups with at least 3 students in each group for a project.

   a. Write an inequality to describe the possible numbers of students in each group.

   b. Use the Multiplication Property of Inequality to write an inequality to describe the possible numbers of students in the class.

**Describe all numbers that satisfy *both* inequalities. Include a graph with your description.**

12. $-3x < -9$ and $-3x < 9$

13. $\dfrac{2}{3}t \geq 4$ and $-\dfrac{3}{5}t \geq -6$

14. $3x < 12$ and $-3x < -3$

15. $\dfrac{y}{5} \leq -2$ and $-\dfrac{y}{4} \geq 1$

# 4.7 Solving Two-Step Inequalities
**For use with Exploration 4.7**

**Learning Target:** Write and solve two-step inequalities.

**Success Criteria:**
- I can apply properties of inequality to generate equivalent inequalities.
- I can solve two-step inequalities using the basic operations.
- I can apply two-step inequalities to solve real-life problems.

---

**1 EXPLORATION: Using Algebra Tiles to Solve Inequalities**

**Work with a partner.**

**a.** What is being modeled by the algebra tiles below? What is the solution?

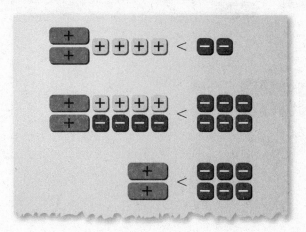

**b.** Use properties of inequality to solve the original inequality in part (a). How do your steps compare to the steps performed with algebra tiles?

---

**4.7**   **Solving Two-Step Inequalities** (continued)

**c.** Write the three inequalities modeled by the algebra tiles below. Then solve each inequality using algebra tiles. Check your answer using properties of inequality.

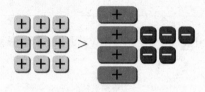

**d.** Explain how solving a two-step inequality is similar to solving a two-step equation.

**Notetaking with Vocabulary**

**Vocabulary:**

**Notes:**

**4.7** **Self-Assessment**

**Use the scale below to rate your understanding of the learning target and the success criteria.**

| 1 | 2 | 3 | 4 |
|---|---|---|---|
| I do not understand. | I can do it with help. | I can do it on my own. | I can teach someone else. |

|  | Rating | Date |
|---|---|---|
| **4.7 Solving Two-Step Inequalities** | | |
| **Learning Target:** Write and solve two-step inequalities. | 1  2  3  4 | |
| I can apply properties of inequality to generate equivalent inequalities. | 1  2  3  4 | |
| I can solve two-step inequalities using the basic operations. | 1  2  3  4 | |
| I can apply two-step inequalities to solve real-life problems. | 1  2  3  4 | |

Name_____  Date_____

**Solve the inequality. Graph the solution.**

**1.** $2 - \frac{q}{3} > 6$

**2.** $7 \le 0.5v + 10$

**3.** $-5(m - 2) > 30$

**4.** $-\frac{3}{2}(f + 6) \le -6$

**5.** $10.5 < 1.5(p - 3)$

**6.** $30 \ge -7.5(w - 4.2)$

**7.** An RV park receives $300 per month from each residential site that is occupied as well as $2000 per month from their overnight sites. Write and solve an inequality to find the number of residential sites that must be occupied to make at least $14,000 in revenue each month.

**8.** Write and solve an inequality that represents the values of $x$ for which the area of the rectangle will be at most 45 square meters.

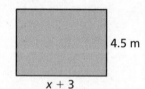

4.5 m

$x + 3$

**Solve the inequality. Graph the solution.**

**9.** $12x - 5x - 4 \ge 60 - 8$

**10.** $4v + 6v + 3.2 < 6.8 - 9.2$

**11.** An animal shelter has fixed weekly expenses of $750. Each animal in the shelter costs an additional $6 a week.

  **a.** During the summer months, the total weekly expenses are at least $1170. Write and solve an inequality that represents the number of animals at the shelter for expenses to be at least $1170 a week.

  **b.** During the winter months, the total weekly expenses are at most $900. Write and solve an inequality that represents the number of animals at the shelter for expenses to be at most $900 a week.

  **c.** The cost for each animal has increased by $2. What will be the maximum weekly expenses during the winter month?

**12.** For what values of $r$ will the area of the shaded region be greater than or equal to half of the area of the larger rectangle?

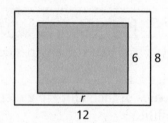

6   8

$r$

12

Name_____ Date_____

# Chapter Self-Assessment

Use the scale below to rate your understanding of the learning target and the success criteria.

| **1** | **2** | **3** | **4** |
|---|---|---|---|
| I do not understand. | I can do it with help. | I can do it on my own. | I can teach someone else. |

|  | Rating | Date |
|---|---|---|
| **4.1 Solving Equations Using Addition or Subtraction** | | |
| **Learning Target:** Write and solve equations using addition or subtraction. | 1  2  3  4 | |
| I can apply the Addition and Subtraction Properties of Equality to produce equivalent equations. | 1  2  3  4 | |
| I can solve equations using addition or subtraction. | 1  2  3  4 | |
| I can apply equations involving addition or subtraction to solve real-life problems. | 1  2  3  4 | |
| **4.2 Solving Equations Using Multiplication or Division** | | |
| **Learning Target:** Write and solve equations using multiplication or division. | 1  2  3  4 | |
| I can apply the Multiplication and Division Properties of Equality to produce equivalent equations. | 1  2  3  4 | |
| I can solve equations using multiplication or division. | 1  2  3  4 | |
| I can apply equations involving multiplication or division to solve real-life problems. | 1  2  3  4 | |
| **4.3 Solving Two-Step Equations** | | |
| **Learning Target:** Write and solve two-step equations. | 1  2  3  4 | |
| I can apply properties of equality to produce equivalent equations. | 1  2  3  4 | |
| I can solve two-step equations using the basic operations. | 1  2  3  4 | |
| I can apply two-step equations to solve real-life problems. | 1  2  3  4 | |

**Chapter 4**

# Chapter Self-Assessment (continued)

|  | Rating | Date |
|---|---|---|
| **4.4 Writing and Graphing Inequalities** | | |
| **Learning Target:** Write inequalities and represent solutions of inequalities on number lines. | 1  2  3  4 | |
| I can write word sentences as inequalities. | 1  2  3  4 | |
| I can determine whether a value is a solution of an inequality. | 1  2  3  4 | |
| I can graph the solutions of inequalities. | 1  2  3  4 | |
| **4.5 Solving Inequalities Using Addition or Subtraction** | | |
| **Learning Target:** Write and solve inequalities using addition or subtraction. | 1  2  3  4 | |
| I can apply the Addition and Subtraction Properties of Inequality to produce equivalent inequalities. | 1  2  3  4 | |
| I can solve inequalities using addition or subtraction. | 1  2  3  4 | |
| I can apply inequalities involving addition or subtraction to solve real-life problems. | 1  2  3  4 | |
| **4.6 Solving Inequalities Using Multiplication or Division** | | |
| **Learning Target:** Write and solve inequalities using multiplication or division. | 1  2  3  4 | |
| I can apply the Multiplication and Division Properties of Inequality to produce equivalent inequalities. | 1  2  3  4 | |
| I can solve inequalities using multiplication or division. | 1  2  3  4 | |
| I can apply inequalities involving multiplication or division to solve real-life problems. | 1  2  3  4 | |
| **4.7 Solving Two-Step Inequalities** | | |
| **Learning Target:** Write and solve two-step inequalities. | 1  2  3  4 | |
| I can apply properties of inequality to generate equivalent inequalities. | 1  2  3  4 | |
| I can solve two-step inequalities using the basic operations. | 1  2  3  4 | |
| I can apply two-step inequalities to solve real-life problems. | 1  2  3  4 | |

**Chapter 5** · **Review & Refresh**

**Simplify.**

1. $\dfrac{3}{18}$

2. $\dfrac{4}{6}$

3. $\dfrac{12}{60}$

4. $\dfrac{14}{28}$

5. $\dfrac{16}{36}$

6. $\dfrac{40}{50}$

**Are the fractions equivalent?**

7. $\dfrac{3}{8} \overset{?}{=} \dfrac{6}{11}$

8. $\dfrac{4}{10} \overset{?}{=} \dfrac{16}{40}$

9. $\dfrac{22}{32} \overset{?}{=} \dfrac{11}{16}$

10. $\dfrac{63}{72} \overset{?}{=} \dfrac{7}{9}$

11. You see 58 birds while on a bird watching tour. Of those birds, you see 12 hawks. Write and simplify the fraction of birds you see that are hawks.

## Chapter 5 Review & Refresh (continued)

**Solve the equation. Check your solution.**

**12.** $\dfrac{d}{12} = -4$

**13.** $-7 = \dfrac{x}{-3}$

**14.** $\dfrac{1}{8}n = 5$

**15.** $6a = -54$

**16.** $10 = -2k$

**17.** $2.7 = -0.9y$

**18.** $-23.4 = -13w$

**19.** $\dfrac{1}{15}z = 6$

**20.** You and three friends spend \$35 on tickets at the movies. Write and solve an equation to find the price $p$ of one ticket.

# Ratios and Ratio Tables
**For use with Exploration 5.1**

**Learning Target:** Understand ratios of rational numbers and use ratio tables to represent equivalent ratios.

**Success Criteria:**
- I can write and interpret ratios involving rational numbers.
- I can use various operations to create tables of equivalent ratios.
- I can use ratio tables to solve ratio problems.

---

**1** **EXPLORATION:** Describing Ratio Relationships

**Work with a partner. Use the recipe shown.**

| Chicken Soup | | | |
|---|---|---|---|
| stewed tomatoes | 9 ounces | chopped spinach | 9 ounces |
| chicken broth | 15 ounces | grated parmesan | 5 tablespoons |
| chopped chicken | 1 cup | | |

**a.** Identify several ratios in the recipe.

**b.** You halve the recipe. Describe your ratio relationships in part (a) using the new quantities. Is the relationship between the ingredients the same as in part (a)? Explain.

**5.1** Ratios and Ratio Tables (continued)

---

**2** **EXPLORATION:** Completing Ratio Tables

**Work with a partner. Use the ratio tables shown.**

| x | 5 | | | |
|---|---|---|---|---|
| y | 1 | | | |

| x | $\frac{1}{4}$ | | | |
|---|---|---|---|---|
| y | $\frac{1}{2}$ | | | |

a. Complete the first ratio table using multiple operations. Use the same operations to complete the second ratio table.

b. Are the ratios in the first table equivalent? the second table? Explain.

c. Do the strategies for completing ratio tables of whole numbers work for completing ratio tables of fractions? Explain your reasoning.

Name_____ Date_____

 **Notetaking with Vocabulary**

**Vocabulary:**

**Notes:**

**5.1 Self-Assessment**

**Use the scale below to rate your understanding of the learning target and the success criteria.**

| **1** | **2** | **3** | **4** |
|---|---|---|---|
| I do not understand. | I can do it with help. | I can do it on my own. | I can teach someone else. |

|  | Rating | Date |
|---|---|---|
| **5.1 Ratios and Ratio Tables** | | |
| **Learning Target:** Understand ratios of rational numbers and use ratio tables to represent equivalent ratios. | 1  2  3  4 | |
| I can write and interpret ratios involving rational numbers. | 1  2  3  4 | |
| I can use various operations to create tables of equivalent ratios. | 1  2  3  4 | |
| I can use ratio tables to solve ratio problems. | 1  2  3  4 | |

## 5.1 Practice

**Write the ratio as a fraction in simplest form.**

1. 198 women to 110 men

2. 1000 songs : 2 megabytes

3. 26.1 miles : 3.6 hours

4. 12 completions to 28 attempts

**Find the missing values in the ratio table. Then write the equivalent ratios.**

5.

| Coffee (ounces) | 4 | 8 | | 16 |
|---|---|---|---|---|
| Cream (ounces) | $\frac{3}{8}$ | | $1\frac{1}{8}$ | |

6.

| Yards | 4 | | | 25 |
|---|---|---|---|---|
| Seconds | 6.4 | 10 | 15.6 | |

7. A water pipe leak floods 50 square feet every $\frac{1}{5}$ minute. What is the flooded area after 6 minutes?

8. You use $\frac{1}{32}$ milliliter of essential oils for every $\frac{3}{4}$ milliliter of avocado oil to make 150 milliliters of a solution.

   a. How much essential oils do you use? How much avocado oil do you use?

   b. You decide that you want to add coconut oil to the solution. You make the new mixture by adding $\frac{3}{16}$ milliliter of coconut oil for every $\frac{1}{32}$ milliliter of essential oils and $\frac{3}{4}$ milliliter of avocado oil. How much essential oils, avocado oil, and coconut oil do you use to make $15\frac{1}{2}$ milliliters of the new solution?

9. The ratio of white rice to water is 1 to $2\frac{1}{2}$. You have $\frac{2}{3}$ cup of rice. How much water do you need?

## 5.2 Rates and Unit Rates
### For use with Exploration 5.2

**Learning Target:** Understand rates involving fractions and use unit rates to solve problems.

**Success Criteria:**
- I can find unit rates for rates involving fractions.
- I can use unit rates to solve rate problems.

---

**1 EXPLORATION:** Writing Rates

**Work with a partner.**

**a.** How many degrees does the minute hand on a clock move every 15 minutes? Write a rate that compares the number of degrees moved by the minute hand to the number of hours elapsed.

**b.** Can you use the rate in part (a) to determine how many degrees the minute hand moves in $\frac{1}{2}$ hour? Explain your reasoning.

---

## 5.2 Rates and Unit Rates (continued)

**c.** Write a rate that represents the number of degrees moved by the minute hand every hour. How can you use this rate to find the number of degrees moved by the minute hand in $2\frac{1}{2}$ hours?

**d.** Draw a clock with hour and minute hands. Draw another clock that shows the time after the minute hand moves 900°. How many degrees does the hour hand move in this time? in one hour? Explain your reasoning.

**Notetaking with Vocabulary**

**Vocabulary:**

**Notes:**

**5.2** **Self-Assessment**

Use the scale below to rate your understanding of the learning target and the success criteria.

| 1 | 2 | 3 | 4 |
|---|---|---|---|
| I do not understand. | I can do it with help. | I can do it on my own. | I can teach someone else. |

|  | Rating | Date |
|---|---|---|
| **5.2 Rates and Unit Rates** | | |
| **Learning Target:** Understand rates involving fractions and use unit rates to solve problems. | 1   2   3   4 | |
| I can find unit rates for rates involving fractions. | 1   2   3   4 | |
| I can use unit rates to solve rate problems. | 1   2   3   4 | |

## 5.2 Practice

**Find the unit rate.**

1. $5.40 for 24 cans

2. $1.29 for 20 ounces

3. 50 meters in 27.5 seconds

4. Find the missing values in the ratio table. Then write the equivalent ratios.

| Distance (Centimeters) | 14 | $\frac{3}{2}$ | | |
|---|---|---|---|---|
| Time (minute) | $\frac{4}{3}$ | | 1 | $\frac{1}{6}$ |

5. There are 16 bacteria in a beaker. Four hours later there are 228 bacteria in the beaker. What is the rate of change per hour in the number of bacteria?

6. The table shows nutritional information for three energy bars.

   a. Which has the most protein per calorie?

   b. Which has the least sugar per calorie?

   c. Which has the highest rate of sugar to fiber?

| Energy Bar | Calories | Protein | Fiber | Sugar |
|---|---|---|---|---|
| A | 220 | 20 g | 12 g | 14 g |
| B | 130 | 12 g | 8 g | 10 g |
| C | 140 | 4 g | 9 g | 9 g |

   d. Compare Bar A with Bar B. Which nutritional item do you think has the highest ratio: calories, protein, fiber, or sugar?

   e. Calculate the ratios in part (d). Which one has the highest ratio?

7. The graph shows the cost of buying scoops of gelato.

   a. What does the point (4, 6) represent?

   b. What is the unit cost?

   c. What is the cost of 12 scoops?

   d. Explain how the graph would change if the unit rate was $1.75 per scoop.

   e. How would the coordinates of the point in part (a) change if the unit rate was $1.75 per scoop?

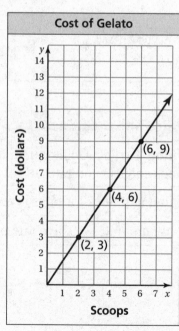

**Cost of Gelato**

# 5.3 Identifying Proportional Relationships
**For use with Exploration 5.3**

**Learning Target:** Determine whether two quantities are in a proportional relationship.

**Success Criteria:**
- I can determine whether ratios form a proportion.
- I can explain how to determine whether quantities are proportional.
- I can distinguish between proportional and nonproportional situations.

---

**1 EXPLORATION: Determining Proportional Relationships**

**Work with a partner.**

**a.** You can paint 50 square feet of a surface every 40 minutes. How long does it take you to paint the mural shown? Explain how you found your answer.

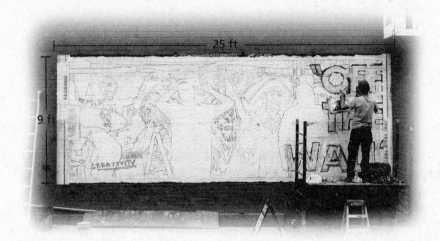

25 ft

9 ft

**b.** The number of square feet you paint is *proportional* to the number of minutes it takes you. What do you think it means for a quantity to be *proportional* to another quantity?

**c.** Assume your friends paint at the same rate as you. The table shows how long it takes you and different numbers of friends to paint a fence. Is $x$ proportional to $y$ in the table? Explain.

| Painters, $x$ | 1 | 2 | 3 | 4 |
|---|---|---|---|---|
| Hours, $y$ | 4 | 2 | $\dfrac{4}{3}$ | 1 |

**d.** How long will it take you and four friends to paint the fence? Explain how you found your answer.

**Notetaking with Vocabulary**

**Vocabulary:**

**Notes:**

**5.3** **Self-Assessment**

Use the scale below to rate your understanding of the learning target and the success criteria.

| 1 | 2 | 3 | 4 |
|---|---|---|---|
| I do not understand. | I can do it with help. | I can do it on my own. | I can teach someone else. |

|  | Rating | Date |
|---|---|---|
| **5.3 Identifying Proportional Relationships** | | |
| **Learning Target:** Determine whether two quantities are in a proportional relationship. | 1  2  3  4 | |
| I can determine whether ratios form a proportion. | 1  2  3  4 | |
| I can explain how to determine whether quantities are proportional. | 1  2  3  4 | |
| I can distinguish between proportional and nonproportional situations. | 1  2  3  4 | |

## 5.3 Practice

**Tell whether the ratios form a proportion.**

**1.** 28.5 : 42 and 19 : 28    **2.** 3.5 : 4 and 11.9 : 13.6    **3.** 124 to 98 and 315 to 225

**Tell whether the rates form a proportion.**

**4.** $24 for 16 burgers; $15 for 10 burgers

**5.** 10 used books for $4.50; 15 used books for $6.00

**6.** The ratio of pennies : nickels is proportional to the ratio of nickels : dimes, and to the ratio of dimes : quarters. If you have one penny and two nickels, how much money do you have?

**7.** The seventh-grade band has 15 drummers and 12 trumpet players. The eighth-grade band has 10 drummers and 8 trumpet players. Do the ratios form a proportion? Explain.

**8.** One mixture contains 6 fluid ounces of water and 10 fluid ounces of vinegar. A second mixture contains 9 fluid ounces of water and 12 fluid ounces of vinegar. Are the mixtures proportional? If not, how much water or vinegar would you add to the second mixture so that they are proportional?

**9.** A wholesale warehouse buys pairs of sandals to sell.

   **a.** The warehouse can purchase 5 pairs of sandals for $65. What is the cost rate?

   **b.** The warehouse can purchase 8 pairs of sandals for $96. What is the cost rate?

   **c.** The warehouse can purchase 10 pairs of sandals for $126.50 and will get one free pair. What is the cost rate?

   **d.** Are any of the cost rates proportional? Explain.

   **e.** Your buyer is to purchase 40 pairs of sandals. Use any combination of parts (a), (b), and (c) for your buyer to purchase the 40 pairs of sandals at the lowest possible cost.

**Find the values of x and y.**

**10.** If $\dfrac{3}{7} = \dfrac{9}{21}$, then $\dfrac{3}{7+x} = \dfrac{9}{21+y}$.

**11.** If $\dfrac{16}{12} = \dfrac{20}{15}$, then $\dfrac{16}{12+x} = \dfrac{20}{15+y}$.

## 5.4 Writing and Solving Proportions
**For use with Exploration 5.4**

**Learning Target:** Use proportions to solve ratio problems.

**Success Criteria:**
- I can solve proportions using various methods.
- I can find a missing value that makes two ratios equivalent.
- I can use proportions to represent and solve real-life problems.

---

**1** **EXPLORATION:** Solving a Ratio Problem

**Work with a partner. A train travels 50 miles every 40 minutes. To determine the number of miles the train travels in 90 minutes, your friend creates the following table.**

| Miles | 50 | $x$ |
|---|---|---|
| Minutes | 40 | 90 |

**a.** Explain how you can find the value of $x$.

**b.** Can you use the information in the table to write a proportion? If so, explain how you can use the proportion to find the value of $x$. If not, explain why not.

## 5.4   Writing and Solving Proportions (continued)

**c.** If a train travels 30 miles every half hour. How far does the train travel in 2 hours?

**d.** Share your results in part (c) with other groups. Compare and contrast methods used to solve the problem.

**Notetaking with Vocabulary**

**Vocabulary:**

**Notes:**

## 5.4 Self-Assessment

**Use the scale below to rate your understanding of the learning target and the success criteria.**

| 1 | 2 | 3 | 4 |
|---|---|---|---|
| I do not understand. | I can do it with help. | I can do it on my own. | I can teach someone else. |

| | Rating | Date |
|---|---|---|
| **5.4 Writing and Solving Proportions** | | |
| **Learning Target:** Use proportions to solve ratio problems. | 1   2   3   4 | |
| I can solve proportions using various methods. | 1   2   3   4 | |
| I can find a missing value that makes two ratios equivalent. | 1   2   3   4 | |
| I can use proportions to represent and solve real-life problems. | 1   2   3   4 | |

Name _____  Date _____

## 5.4  Practice

1. Describe and correct the error in writing the proportion.

2. There are 3 referees for every 16 players. Write a proportion that gives the number of referees $r$ for 128 players.

|   | Day 1 | Day 2 |
|---|---|---|
| **Length** | 3.1 | 15.5 |
| **Height** | $h$ | 45 |

$$\frac{15.5}{h} = \frac{3.1}{45}$$

**Solve the proportion.**

3. $\dfrac{5}{12} = \dfrac{x}{36}$

4. $\dfrac{20}{3.4} = \dfrac{800}{y}$

5. $\dfrac{2.8}{r} = \dfrac{70}{3}$

6. You can buy 48 seashells for $15. How many seashells can you buy for $37.50?

7. You need 252 gift bags for an event. You found a gift bag that costs $6.50 for 21 bags. Your budget is $75. Is the gift bag within your budget? Justify your answer.

8. A recipe calls for $\frac{3}{4}$ cup of sugar and $\frac{1}{2}$ cup of brown sugar. You are reducing the recipe. You will use $\frac{1}{6}$ cup of brown sugar. How much sugar will you use?

9. A calculator has 50 keys in five colors: gray, black, blue, yellow, and green.

   a. There are 6 gray keys for every 7 blue keys. Write the possible ratios for gray to blue keys.

   b. There are 6 gray keys for every 11 black keys. Write the possible ratios for gray to black keys.

   c. There are 6 gray keys for every 11 black keys. Also, the number of black keys is 2 less than twice the number of gray keys. Use your answer to part (b) to determine how many gray keys and how many black keys there are.

   d. There is 1 yellow key for every 1 green key. How many keys of each color are there?

10. Give two possible pairs of values for $p$ and $q$: $\dfrac{7}{10} = \dfrac{p}{q}$.

## 5.5 Graphs of Proportional Relationships
**For use with Exploration 5.5**

**Learning Target:** Represent proportional relationships using graphs and equations.

**Success Criteria:**
- I can determine whether quantities are proportional using a graph.
- I can find the unit rate of a proportional relationship using a graph.
- I can create equations to represent proportional relationships.

---

**1  EXPLORATION: Representing Relationships Graphically**

**Work with a partner. The tables represent two different ways that red and blue food coloring are mixed.**

| Mixture 1 | |
|---|---|
| Drops of Blue, x | Drops of Red, y |
| 1 | 2 |
| 2 | 4 |
| 3 | 6 |
| 4 | 8 |

| Mixture 2 | |
|---|---|
| Drops of Blue, x | Drops of Red, y |
| 0 | 2 |
| 2 | 4 |
| 4 | 6 |
| 6 | 8 |

**a.** Represent each table in the same coordinate plane. Which graph represents a proportional relationship? How do you know?

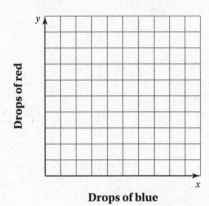

**5.5** **Graphs of Proportional Relationships** (continued)

**b.** Find the unit rate of the proportional relationship. How is the unit rate shown on the graph?

**c.** What is the multiplicative relationship between $x$ and $y$ for the proportional relationship? How can you use this value to write an equation that relates $y$ and $x$?

Name_____ Date_____

## 5.5 Notetaking with Vocabulary

**Vocabulary:**

**Notes:**

## 5.5 Self-Assessment

Use the scale below to rate your understanding of the learning target and the success criteria.

| **1** | **2** | **3** | **4** |
|---|---|---|---|
| I do not understand. | I can do it with help. | I can do it on my own. | I can teach someone else. |

| | Rating | Date |
|---|:---:|:---:|
| **5.5 Graphs of Proportional Relationships** | | |
| **Learning Target:** Represent proportional relationships using graphs and equations. | 1   2   3   4 | |
| I can determine whether quantities are proportional using a graph. | 1   2   3   4 | |
| I can find the unit rate of a proportional relationship using a graph. | 1   2   3   4 | |
| I can create equations to represent proportional relationships. | 1   2   3   4 | |

## 5.5 Practice

**Interpret each plotted point in the graph of the proportional relationship. Then identify the unit rate.**

**1.**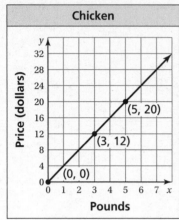

**2.**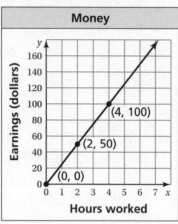

**The graph of a proportional relationship passes through the given points. Find y.**

**3.** $(4, 8), (1, y)$

**4.** $(3, 21), (1, y)$

**5.** $(1.5, 9), (1, y)$

**6.** $(3.5, 14), (1, y)$

**7.** Two classes have car washes to raise money for class trips. A portion of the earnings will pay for using the two locations for the car washes. The graph shows that the trip earnings of the two classes are proportional to the car wash earnings.

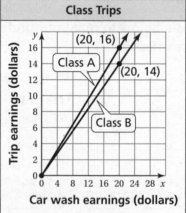

   **a.** Express the trip earnings rate for each class as a percent.

   **b.** What trip earnings does Class A receive for earning $75 from the car wash?

   **c.** How much less does Class B receive than Class A for earning $75 from the car wash?

**The variables x and y are proportional. Use the values to find the constant of proportionality. Then write an equation that relates x and y.**

**8.** When $y = 36$, $x = 18$.    **9.** When $y = 51$, $x = 34$.    **10.** When $y = 55$, $x = 10$.

**11.** Does the graph of every proportional relationship pass through the origin? Is every relationship whose graph passes through the origin a proportional relationship? Explain your reasoning.

## 5.6 Scale Drawings
**For use with Exploration 5.6**

**Learning Target:** Solve problems involving scale drawings.

**Success Criteria:**
- I can find an actual distance in a scale drawing.
- I can explain the meaning of scale and scale factor.
- I can use a scale drawing to find the actual lengths and areas of real-life objects.

---

**1  EXPLORATION: Creating a Scale Drawing**

**Work with a partner. Several sections in a zoo are drawn on 1-centimeter grid paper as shown. Each centimeter in the drawing represents 4 meters.**

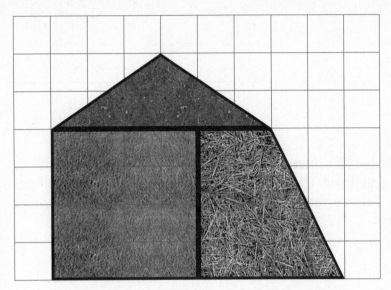

a. Describe the relationship between the lengths of the fences in the drawing and the actual side lengths of the fences.

## 5.6 Scale Drawings (continued)

**b.** Describe the relationship between the areas of the sections in the drawing and the actual areas of the sections.

**c.** Are the relationships in parts (a) and (b) the same? Explain your reasoning.

**d.** Choose a different distance to represent each centimeter on a piece of 1-centimeter grid paper. Then create a new drawing of the sections in the zoo using the distance you chose. Describe any similarities or differences in the drawings.

## 5.6 Notetaking with Vocabulary

**Vocabulary:**

**Notes:**

## 5.6 Self-Assessment

**Use the scale below to rate your understanding of the learning target and the success criteria.**

| 1 | 2 | 3 | 4 |
|---|---|---|---|
| I do not understand. | I can do it with help. | I can do it on my own. | I can teach someone else. |

| | Rating | Date |
|---|---|---|
| **5.6 Scale Drawings** | | |
| **Learning Target:** Solve problems involving scale drawings. | 1  2  3  4 | |
| I can find an actual distance in a scale drawing. | 1  2  3  4 | |
| I can explain the meaning of scale and scale factor. | 1  2  3  4 | |
| I can use a scale drawing to find the actual lengths and areas of real-life objects. | 1  2  3  4 | |

**Big Ideas Math: Modeling Real Life Grade 7 Accelerated** **125**
Student Journal

# 5.6 Practice

1. In the actual blueprint of the bedroom suite, each square has a side length of $\frac{1}{2}$ inch.

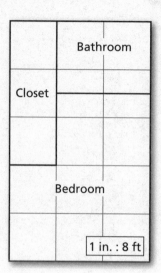

   a. What are the dimensions of the bedroom suite?

   b. What are the dimensions of the bathroom?

   c. What is the length of the longest wall in the bedroom?

   d. What is the ratio of the perimeter of the closet to the perimeter of the bathroom?

   e. What is the ratio of the area of the closet to the area of the bathroom? How can you explain this by looking at the squares in each?

   f. All of the walls in the bedroom suite are covered with drywall. Which will cost the most to drywall – *the closet, the bathroom,* or *both are the same*?

   g. All of the floors in the bedroom suite are covered with the same tile. Which will cost the most to tile – *the closet, the bathroom* or *both are the same*?

**Find the missing dimension. Use the scale 2 : 5.**

| Model | Actual |
|---|---|
| **2.** Depth: 10 km | Depth: ____ km |
| **3.** Length: 5 in. | Length: ____ in. |
| **4.** Length: ____ ft | Length: 24 ft |
| **5.** Diameter: ____ m | Diameter: 32.5 m |

6. A scale model has a scale of 1 ft : 8 ft. Describe and correct the error in finding the model length that corresponds to the actual length 48 feet.

   $$\cancel{\quad} \quad \frac{1}{8} = \frac{48 \text{ ft}}{x \text{ ft}}$$
   $$x = 384 \text{ ft}$$

Name_____ Date_____

# Chapter Self-Assessment

Use the scale below to rate your understanding of the learning target and the success criteria.

**1** I do not understand.  **2** I can do it with help.  **3** I can do it on my own.  **4** I can teach someone else.

|  | Rating | Date |
|---|---|---|
| **5.1 Ratios and Ratio Tables** | | |
| **Learning Target:** Understand ratios of rational numbers and use ratio tables to represent equivalent ratios. | 1  2  3  4 | |
| I can write and interpret ratios involving rational numbers. | 1  2  3  4 | |
| I can use various operations to create tables of equivalent ratios. | 1  2  3  4 | |
| I can use ratio tables to solve ratio problems. | 1  2  3  4 | |
| **5.2 Rates and Unit Rates** | | |
| **Learning Target:** Understand rates involving fractions and use unit rates to solve problems. | 1  2  3  4 | |
| I can find unit rates for rates involving fractions. | 1  2  3  4 | |
| I can use unit rates to solve rate problems. | 1  2  3  4 | |
| **5.3 Identifying Proportional Relationships** | | |
| **Learning Target:** Determine whether two quantities are in a proportional relationship. | 1  2  3  4 | |
| I can determine whether ratios form a proportion. | 1  2  3  4 | |
| I can explain how to determine whether quantities are proportional. | 1  2  3  4 | |
| I can distinguish between proportional and nonproportional situations. | 1  2  3  4 | |

Name _____ Date _____

| | Rating | Date |
|---|---|---|
| **5.4 Writing and Solving Proportions** | | |
| **Learning Target:** Use proportions to solve ratio problems. | 1   2   3   4 | |
| I can solve proportions using various methods. | 1   2   3   4 | |
| I can find a missing value that makes two ratios equivalent. | 1   2   3   4 | |
| I can use proportions to represent and solve real-life problems. | 1   2   3   4 | |
| **5.5 Graphs of Proportional Relationships** | | |
| **Learning Target:** Represent proportional relationships using graphs and equations. | 1   2   3   4 | |
| I can determine whether quantities are proportional using a graph. | 1   2   3   4 | |
| I can find the unit rate of a proportional relationship using a graph. | 1   2   3   4 | |
| I can create equations to represent proportional relationships. | 1   2   3   4 | |
| **5.6 Scale Drawings** | | |
| **Learning Target:** Solve problems involving scale drawings. | 1   2   3   4 | |
| I can find an actual distance in a scale drawing. | 1   2   3   4 | |
| I can explain the meaning of scale and scale factor. | 1   2   3   4 | |
| I can use a scale drawing to find the actual lengths and areas of real-life objects. | 1   2   3   4 | |

Name_____ Date_____

**Write the percent as a fraction or mixed number in simplest form.**

**1.** 25%

**2.** 65%

**3.** 110%

**4.** 250%

**5.** 15%

**6.** 6%

**7.** A store marks up a pair of sneakers 30%. Write the percent as a fraction or mixed number in simplest form.

## Chapter 6 Review & Refresh (continued)

**Write the fraction or mixed number as a percent.**

8. $\dfrac{1}{5}$

9. $\dfrac{1}{4}$

10. $\dfrac{21}{25}$

11. $1\dfrac{2}{5}$

12. $2\dfrac{13}{20}$

13. $1\dfrac{1}{2}$

14. You own $\dfrac{3}{5}$ of the coins in a collection. What percent of the coins do you own?

## 6.1 Fractions, Decimals, and Percents
**For use with Exploration 6.1**

**Learning Target:** Rewrite fractions, decimals, and percents using different representations.

**Success Criteria:**
- I can write percents as decimals and decimals as percents.
- I can write fractions as decimals and percents.
- I can compare and order fractions, decimals, and percents.

---

**1 EXPLORATION:** Comparing Numbers in Different Forms

**Work with a partner. Determine which number is greater. Explain your method.**

**a.** 7% sales tax   or   $\frac{1}{20}$ sales tax

**b.** 0.37 cup of flour   or   $\frac{1}{3}$ cup of flour

**c.** $\frac{5}{8}$-inch wrench   or   0.375-inch wrench

**d.** $12\frac{3}{5}$ dollars   or   12.56 dollars

**e.** $5\frac{5}{6}$ fluid ounces   or   5.6 fluid ounces

---

**6.1** **Fractions, Decimals, and Percents** (continued)

## 2 **EXPLORATION:** Ordering Fractions, Decimals, and Percents

**Work with a partner and follow the steps below.**

- Write five different numbers on individual slips of paper. Include at least one decimal, one fraction, and one percent.

- On a separate sheet of paper, create an answer key that shows your numbers written from least to greatest.

- Exchange slips of paper with another group and race to order the numbers from least to greatest. Then exchange answer keys to check your orders.

##  Notetaking with Vocabulary

**Vocabulary:**

**Notes:**

## 6.1 Self-Assessment

Use the scale below to rate your understanding of the learning target and the success criteria.

| **1** | **2** | **3** | **4** |
|---|---|---|---|
| I do not understand. | I can do it with help. | I can do it on my own. | I can teach someone else. |

|  | Rating | Date |
|---|---|---|
| **6.1 Fractions, Decimals, and Percents** | | |
| **Learning Target:** Rewrite fractions, decimals, and percents using different representations. | 1   2   3   4 | |
| I can write percents as decimals and decimals as percents. | 1   2   3   4 | |
| I can write fractions as decimals and percents. | 1   2   3   4 | |
| I can compare and order fractions, decimals, and percents. | 1   2   3   4 | |

## 6.1 Practice

**Determine which number is greater. Explain your method.**

1. $\frac{1}{4}$, 22%

2. $\frac{5}{9}$, 55%

3. 3.2, 32%

4. 99.9%, 1

**Write the fraction as a decimal and a percent.**

5. $\frac{13}{6}$

6. $\frac{25}{11}$

7. $\frac{1}{300}$

8. $\frac{3}{1000}$

**Order the numbers from least to greatest. Explain your method.**

9. $\frac{4}{3}$, 1.333, 133.33%, 1.334

10. 81.8%, 0.8182, $0.8\overline{2}$, $\frac{9}{11}$

11. Your band played for the local elementary school. After the concert, they surveyed the 2nd, 3rd, 4th, and 5th graders as to their interest in playing an instrument. The table gives the results of the survey.

| Grade | 2nd | 3rd | 4th | 5th |
|---|---|---|---|---|
| Portion of Grade Interested in Playing an Instrument | 0.52 | $\frac{25}{36}$ | 61.24% | $0.572\overline{4}$ |

   a. What grade had the greatest portion of students interested in playing an instrument?

   b. What grade had the least portion of students interested in playing an instrument?

   c. How many times more is the portion of students in part (a) than the portion of students in part (b)?

   d. Did the interest in playing an instrument improve as students advanced from grade to grade, yes or no? Explain your answer.

12. At least 52% of the 25 students in a class brought their own lunch last month. What are the possible numbers of students in the class who brought their own lunch last month? Justify your answer.

13. Which do you prefer, writing a decimal as a percent or writing a fraction as a percent. Explain your answer.

## 6.2 The Percent Proportion
### For use with Exploration 6.2

**Learning Target:** Use the percent proportion to find missing quantities.

**Success Criteria:**
- I can write proportions to represent percent problems.
- I can solve a proportion to find a percent, a part, or a whole.

### 1 EXPLORATION: Using Percent Models

**Work with a partner.**

a. Complete each model. Explain what each model represents.

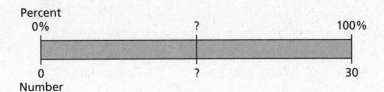

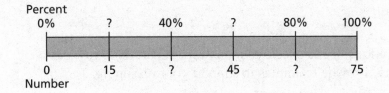

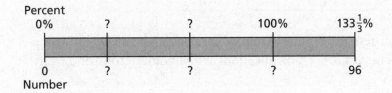

## 6.2 The Percent Proportion (continued)

**b.** Use the models in part (a) to answer each question.

What number is 50% of 30?

15 is what percent of 75?

96 is $133\frac{1}{3}\%$ of what number?

**c.** How can you use ratio tables to check your answers in part (b)? How can you use proportions? Provide examples to support your reasoning.

**d.** Write a question different from those in part (b) that can be answered using one of the models in part (a). Trade questions with another group and find the solution.

## 6.2 Notetaking with Vocabulary

**Vocabulary:**

**Notes:**

## 6.2 Self-Assessment

Use the scale below to rate your understanding of the learning target and the success criteria.

| **1** | **2** | **3** | **4** |
|---|---|---|---|
| I do not understand. | I can do it with help. | I can do it on my own. | I can teach someone else. |

| | Rating | Date |
|---|---|---|
| **6.2 The Percent Proportion** | | |
| **Learning Target:** Use the percent proportion to find missing quantities. | 1   2   3   4 | |
| I can write proportions to represent percent problems. | 1   2   3   4 | |
| I can solve a proportion to find a percent, a part, or a whole. | 1   2   3   4 | |

## 6.2 Practice

**Write and solve a proportion to answer the question.**

1. 55% of what number is 33?

2. What percent of 120 is 42?

3. 36 is 0.8% of what number?

4. 48 is what percent of 64?

5. Of the 360 runners at a 5-kilometer race, 20% are in the 35–39 age bracket. How many runners at the 5-kilometer race are in the 35–39 age bracket?

6. You pay $3.69 for a gallon of gasoline. This is 90% of the price of a gallon of gasoline one year ago. What was the price of a gallon of gasoline one year ago?

7. Describe and correct the error in using the percent proportion to answer the question below.

   "6 is 6.25% of what number?"

$$\times \quad \frac{a}{w} = \frac{p}{100}$$

$$\frac{6}{w} = \frac{0.0625}{100}$$

$$w = 9600$$

**Write and solve a proportion to answer the question.**

8. 70% of the class likes chocolate milk. There are 21 students in the class who like chocolate milk. How many students are in the class?

9. A pizza costs $7.20. This is 250% of the cost of a drink. What is the cost of the drink?

10. A reduced calorie recipe requires 72% of the normal amount of sugar. The normal amount of sugar is $\frac{3}{8}$ cup. How much sugar should you add to the reduced calorie recipe?

11. 1.4 is what percent of 1.12?

12. You earn a score of 86.8 on a standardized exam. Your score is 140% higher than your friend's score on the standardized exam. What is your friend's score?

13. 80% of a number is $x$. What is 40% of the number? Assume $x > 0$.

14. Answer each question. Assume $x > 0$.

    a. What is 35% of $90x$?

    b. What percent of $16x$ is $9x$?

## 6.3 The Percent Equation
### For use with Exploration 6.3

**Learning Target:** Use the percent equation to find missing quantities.

**Success Criteria:**
- I can write equations to represent percent problems.
- I can use the percent equation to find a percent, a part, or a whole.

**1 EXPLORATION: Using Percent Equations**

**Work with a partner.**

a. The circle graph shows the number of votes received by each candidate during a school election. So far, only half of the students have voted. Find the percent of students who voted for each candidate. Explain your method.

**Votes Received by Each Candidate**

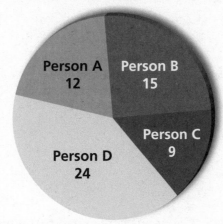

b. You have learned that $\dfrac{\text{part}}{\text{whole}} = \text{percent}$. Solve the equation for the "part".

Explain your reasoning

**6.3**   **The Percent Equation** (continued)

**c.** The circle graph shows the final results
of the election after every student voted.
Use the equation you wrote in part (b) to
find the number of students who voted
for each candidate.

**Final Results**

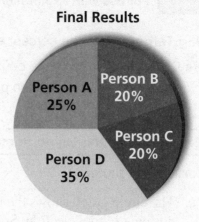

**d.** Use a different method to check your answers in part (c). Which method do
you prefer? Explain.

**Notetaking with Vocabulary**

**Vocabulary:**

**Notes:**

## 6.3  Self-Assessment

Use the scale below to rate your understanding of the learning target and the success criteria.

| 1 | 2 | 3 | 4 |
|---|---|---|---|
| I do not understand. | I can do it with help. | I can do it on my own. | I can teach someone else. |

|  | Rating | Date |
|---|---|---|
| **6.3 The Percent Equation** | | |
| **Learning Target:** Use the percent equation to find missing quantities. | 1  2  3  4 | |
| I can write equations to represent percent problems. | 1  2  3  4 | |
| I can use the percent equation to find a percent, a part, or a whole. | 1  2  3  4 | |

## 6.3 Practice

**Write and solve an equation to answer the question.**

1. 27 is 0.5% of what number?

2. What number is 125% of 240?

3. 1.4% of what number is 28?

4. 27 is what percent of 72?

5. During a given month, there was a total of 23.6 inches of rain. This was 250% of the average rainfall for that month. What is the average rainfall for that month?

6. To maintain an acceptable level of chlorine in your pool, you add 1.4 gallons of chlorine. This is 0.007% of the amount of water in your pool. How many gallons of water are in your pool?

7. You must attend a minimum of 85% of the practices in order to play in the playoffs. You have made 37 of the 42 practices. Will you be able to play in the playoffs?

8. You are in charge of the seventh-grade graduation dinner. The table shows the results of a survey of students' meal preferences.

   | Choice | Percent |
   |---|---|
   | Chicken Nuggets | 25% |
   | Spaghetti | ? |
   | Pizza | 45% |
   | Fish Sticks | ? |

   a. 144 students chose pizza. How many students responded to the survey?

   b. How many students chose chicken nuggets?

   c. The number of students choosing fish sticks was 50% of the number of students choosing spaghetti. How many students chose fish sticks?

   d. How many students chose spaghetti?

9. What is 15% of 40% of $180?

10. There are 15 copies of a popular CD left to be sold in a store. This is between 1% and 1.5% of the original number of copies of the CD in the store. The original number of CDs was between what two numbers?

11. Tell whether the statement is *true* or *false*. Explain your reasoning.

    If $A$ is 45% of $B$, then the ratio of $A : B$ is $9 : 20$.

## 6.4 Percents of Increase and Decrease
**For use with Exploration 6.4**

**Learning Target:** Find percents of change in quantities.

**Success Criteria:**
- I can explain the meaning of percent of change.
- I can find the percent of increase or decrease in a quantity.
- I can find the percent error of a quantity.

### 1 EXPLORATION: Exploring Percent of Change

**Work with a partner.**

Each year in the Columbia River Basin, adult salmon swim upriver to streams to lay eggs.

To go up river, the adult salmon use fish ladders. But to go down river, the young salmon must pass through several dams.

At one time, there were electric turbines at each of the eight dams on the main stem of the Columbia and Snake Rivers. About 88% of the young salmon pass through a single dam unharmed.

a. One thousand young salmon pass through a dam. How many pass through unharmed?

b. One thousand young salmon pass through the river basin. How many pass through all 8 dams unharmed?

**6.4** **Percents of Increase and Decrease** (continued)

    **c.** By what percent does the number of young salmon *decrease* when passing through a single dam?

    **d.** Describe a similar real-life situation in which a quantity *increases* by a constant percent each time an event occurs.

Name_____ Date_____

**Notetaking with Vocabulary**

**Vocabulary:**

**Notes:**

**Self-Assessment**

Use the scale below to rate your understanding of the learning target and the success criteria.

| **1** | **2** | **3** | **4** |
|---|---|---|---|
| I do not understand. | I can do it with help. | I can do it on my own. | I can teach someone else. |

|  | Rating | Date |
|---|---|---|
| **6.4 Percents of Increase and Decrease** | | |
| **Learning Target:** Find percents of change in quantities. | 1  2  3  4 | |
| I can explain the meaning of percent of change. | 1  2  3  4 | |
| I can find the percent of increase or decrease in a quantity. | 1  2  3  4 | |
| I can find the percent error of a quantity. | 1  2  3  4 | |

# 6.4 Practice

**Identify the percent of change as an *increase* or a *decrease*. Then find the percent of change. Round to the nearest tenth of a percent if necessary.**

1. 3.2 kilograms to 2.4 kilograms

2. 41 euros to 85 euros

3. $\frac{2}{7}$ to $\frac{4}{7}$

4. $\frac{5}{6}$ to $\frac{1}{3}$

5. Last month you swam the 50-meter freestyle in 28.38 seconds. Today you swam it in 27.33 seconds. What is your percent of change? Round to the nearest tenth of a percent if necessary.

6. Last week 1200 burgers were served at a restaurant.

   a. This week 1176 burgers were served. What is the percent of change?

   b. Use the percent of change from part (a) to predict the number of burgers served next week. Round to the nearest whole number if necessary.

7. The price of a share of a stock was $37.50 yesterday.

   a. Today there was a price decrease of 4%. What is today's price?

   b. Based on today's price in part (a), what percent of change is needed to bring the price back up to $37.50? Round to the nearest tenth of a percent if necessary.

8. The table shows the membership of two scout troops.

   | Year | Troop A | Troop B |
   |------|---------|---------|
   | 2010 | 14 | 21 |
   | 2011 | 16 | 24 |

   a. What is the percent of change in membership from 2010 to 2011 for Troop A? Round to the nearest tenth of a percent if necessary.

   b. What is the percent of change in membership from 2010 to 2011 for Troop B? Round to the nearest tenth of a percent if necessary.

   c. Which troop has the better record in terms of the number of new members?

   d. Which troop has the better record in terms of the percent of change in membership?

# 6.5 Discounts and Markups
**For use with Exploration 6.5**

**Learning Target:** Solve percent problems involving discounts and markups.

**Success Criteria:**
- I can use percent models to solve problems involving discounts and markups.
- I can write and solve equations to solve problems involving discounts and markups.

---

**1 EXPLORATION: Comparing Discounts**

**Work with a partner.**

a. The same pair of earrings is on sale at three stores. Which store has the best price? Use the percent models to justify your answer.

Store A:
Regular Price: $45

Store B:
Regular Price: $49

Store C:
Regular Price: $39

40% off

50% off

20% off

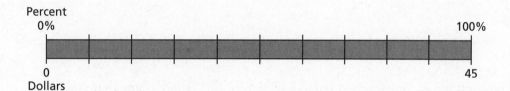

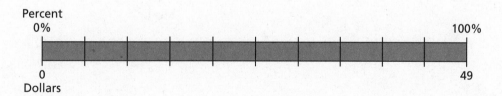

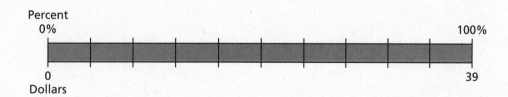

**6.5**  **Discounts and Markups** (continued)

**b.** You buy the earrings on sale for 30% off at a different store. You pay
$22.40. What was the original price of the earrings? Use the percent model
to justify your answer.

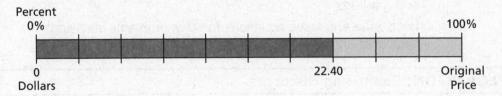

**c.** You sell the earrings in part (b) to a friend for 60% more than what you
paid. What is the selling price? Use a percent model to justify your answer.

## 6.5 Notetaking with Vocabulary

**Vocabulary:**

**Notes:**

## 6.5 Self-Assessment

**Use the scale below to rate your understanding of the learning target and the success criteria.**

| 1 | 2 | 3 | 4 |
|---|---|---|---|
| I do not understand. | I can do it with help. | I can do it on my own. | I can teach someone else. |

| | Rating | Date |
|---|---|---|
| **6.5 Discounts and Markups** | | |
| **Learning Target:** Solve percent problems involving discounts and markups. | 1  2  3  4 | |
| I can use percent models to solve problems involving discounts and markups. | 1  2  3  4 | |
| I can write and solve equations to solve problems involving discounts and markups. | 1  2  3  4 | |

## 6.5 Practice

**Find the original price, discount, sale price, selling price, markup, or cost to store. Round to the nearest penny if necessary.**

**1.** Original price: $130

Discount: 45%

Sale price: ____

**2.** Original price: $500

Discount: ____

Sale price: $175

**3.** Original price: ____

Discount: 5%

Sale price: $68.40

**4.** Cost to store: $1600

Markup: 33%

Selling price: ____

**5.** Cost to store: $65

Markup: ____

Selling price: $91

**6.** Cost to store: ____

Markup: 25%

Selling price: $437.50

**7.** You are buying shoes online. The selling price is $29.99. Round to the nearest penny if necessary.

    **a.** The sales tax is 6.5%. What is the total cost?

    **b.** The cost of shipping is 15% of the total cost. What is the total cost plus shipping?

    **c.** If the total cost plus shipping is greater than $35, then you receive a 10% discount off the original selling price. Do you qualify? If so, what is the new total cost plus shipping?

**8.** You have a coupon for $15 off a video game. You can use it on 2 separate days.

    **a.** On Monday, you buy a video game. The discounted price of your video game is $22.99. What is the original price of the game?

    **b.** What is the percent of the discount to the nearest percent for the video game that you purchased on Monday?

    **c.** On Thursday, you buy another video game. The discounted price of your video game is $12.99. What is the original price of the game?

    **d.** What is the percent of the discount to the nearest percent for the video game that you purchased on Thursday?

## 6.6 Simple Interest
**For use with Exploration 6.6**

**Learning Target:** Understand and apply the simple interest formula.

**Success Criteria:** • I can explain the meaning of simple interest.
• I can use the simple interest formula to solve problems.

**1 EXPLORATION:** Understanding Simple Interest

**Work with a partner. You deposit $150 in an account that earns 6% *simple interest per year*. You do not make any other deposits or withdrawals. The table shows the balance of the account at the end of each year.**

| Years | Balance |
|-------|---------|
| 0 | $150 |
| 1 | $159 |
| 2 | $168 |
| 3 | $177 |
| 4 | $186 |
| 5 | $195 |
| 6 | $204 |

**a.** Describe any patterns you see in the account balance.

**b.** How is the amount of interest determined each year?

**6.6** **Simple Interest** (continued)

**c.** How can you find the amount of simple interest earned when you are given an initial amount, an interest rate, and a period of time?

**d.** You deposit $150 in a different account that earns simple interest. The table shows the balance of the account each year. What is the interest rate of the account? What is the balance after 10 years?

| Years   | 0     | 1     | 2     | 3     |
|---------|-------|-------|-------|-------|
| Balance | $150  | $165  | $180  | $195  |

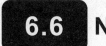

# Notetaking with Vocabulary

**Vocabulary:**

**Notes:**

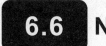

# Self-Assessment

Use the scale below to rate your understanding of the learning target and the success criteria.

| **1** | **2** | **3** | **4** |
|---|---|---|---|
| I do not understand. | I can do it with help. | I can do it on my own. | I can teach someone else. |

| | Rating | Date |
|---|---|---|
| **6.6 Simple Interest** | | |
| **Learning Target:** Understand and apply the simple interest formula. | 1　2　3　4 | |
| I can explain the meaning of simple interest. | 1　2　3　4 | |
| I can use the simple interest formula to solve problems. | 1　2　3　4 | |

## 6.6 Practice

**An account earns simple interest. (a) Find the interest earned. (b) Find the balance of the account.**

**1.** $2600 at 3.2% for 4 years

**2.** $75,000 at 8.5% for 3 months

**Find the annual interest rate or find the amount of time.**

**3.** $I = \$41.80$, $P = \$440$, $t = 2$ years

**4.** $I = \$893.75$, $P = \$5500$, $t = 30$ months

**5.** $I = \$9.90$, $P = \$360$, $r = 5.5\%$

**6.** $I = \$2064$, $P = \$10,000$, $r = 6.88\%$

**Find the amount paid for the loan.**

**7.** You borrow $20,000 to purchase a car. The interest rate is 7.5%. The loan is for 10 years. What is the total amount paid? How much would the payment be each month?

**8.** You invest $6000 in a restaurant. The investment has a return of 12%. How much do you receive for your investment after 30 months?

**9.** You deposit $2000 in an account. The account earns $120 simple interest in 8 months. What is the annual interest rate?

**10.** You put money in two different accounts for one year each. The total simple interest for the two accounts is $140. You earn 6% interest on the first account, in which you deposited $1000. You deposited $800 in the second account. What is the annual interest rate for the second account?

**11.** You deposit $1200 in an account.

   **a.** The account earns 2.7% simple interest per year. What is the balance of the account after 3 months?

   **b.** The interest rate changes, and your new balance now earns 2% simple interest per year. What is the balance of the account after the next 6 months? Round to the nearest penny if necessary.

   **c.** The interest rate changes again, and your new balance now earns 2.6% simple interest per year. What is the balance of the account after an additional 3 months? Round to the nearest penny if necessary.

   **d.** How much did the account earn in simple interest for the year?

   **e.** Based on the interest in part (d), what was the actual simple interest rate for the year? Round to the nearest tenth of a percent.

**12.** You purchase a new guitar and take out a loan for $450. You have 18 equal monthly payments of $28 each. What is the simple interest rate for the loan? Round to the nearest tenth of a percent if necessary.

Name_____ Date_____

# Chapter Self-Assessment

Use the scale below to rate your understanding of the learning target and the success criteria.

| 1 | 2 | 3 | 4 |
|---|---|---|---|
| I do not understand. | I can do it with help. | I can do it on my own. | I can teach someone else. |

| | Rating | Date |
|---|---|---|
| **6.1 Fractions, Decimals, and Percents** | | |
| **Learning Target:** Rewrite fractions, decimals, and percents using different representations. | 1   2   3   4 | |
| I can write percents as decimals and decimals as percents. | 1   2   3   4 | |
| I can write fractions as decimals and percents. | 1   2   3   4 | |
| I can compare and order fractions, decimals, and percents. | 1   2   3   4 | |
| **6.2 The Percent Proportion** | | |
| **Learning Target:** Use the percent proportion to find missing quantities. | 1   2   3   4 | |
| I can write proportions to represent percent problems. | 1   2   3   4 | |
| I can solve a proportion to find a percent, a part, or a whole. | 1   2   3   4 | |
| **6.3 The Percent Equation** | | |
| **Learning Target:** Use the percent equation to find missing quantities. | 1   2   3   4 | |
| I can write equations to represent percent problems. | 1   2   3   4 | |
| I can use the percent equation to find a percent, a part, or a whole. | 1   2   3   4 | |

**Chapter 6**

# Chapter Self-Assessment (continued)

| | Rating | Date |
|---|---|---|
| **6.4 Percents of Increase and Decrease** | | |
| **Learning Target:** Find percents of change in quantities. | 1    2    3    4 | |
| I can explain the meaning of percent of change. | 1    2    3    4 | |
| I can find the percent of increase or decrease in a quantity. | 1    2    3    4 | |
| I can find the percent error of a quantity. | 1    2    3    4 | |
| **6.5 Discounts and Markups** | | |
| **Learning Target:** Solve percent problems involving discounts and markups. | 1    2    3    4 | |
| I can use percent models to solve problems involving discounts and markups. | 1    2    3    4 | |
| I can write and solve equations to solve problems involving discounts and markups. | 1    2    3    4 | |
| **6.6 Simple Interest** | | |
| **Learning Target:** Understand and apply the simple interest formula. | 1    2    3    4 | |
| I can explain the meaning of simple interest. | 1    2    3    4 | |
| I can use the simple interest formula to solve problems. | 1    2    3    4 | |

# Review & Refresh

**Write the ratio in simplest form.**

**1.** bats to baseballs

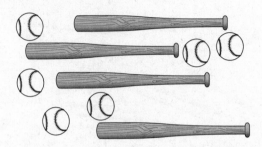

**2.** bows to gift boxes

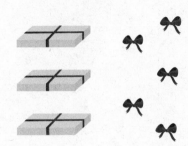

**3.** hammers to screwdrivers

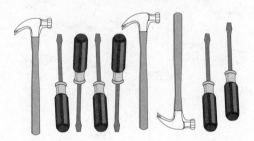

**4.** apples to bananas

**5.** There are 100 students in the sixth grade. There are 15 sixth-grade teachers. What is the ratio of teachers to students?

**Chapter 7 Review & Refresh** (continued)

**Write the ratio in simplest form.**

**6.** golf balls to total number of balls

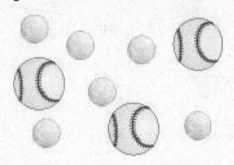

**7.** rulers to total pieces of equipment

**8.** apples to total number of fruit

**9.** small fish to total number of fish

**10.** There are 24 flute players and 18 trumpet players in the band. Write the ratio of trumpet players to total number of trumpet players and flute players.

**7.1** **Probability**
For use with Exploration 7.1

**Learning Target:** Understand how the probability of an event indicates its likelihood.

**Success Criteria:**
- I can identify possible outcomes of an experiment.
- I can use probability and relative frequency to describe the likelihood of an event.
- I can use relative frequency to make predictions.

**1** **EXPLORATION:** Determining Likelihood

**Work with a partner. Use the spinners shown.**

a. For each spinner, determine which numbers you are more likely to spin and which numbers you are less likely to spin. Explain your reasoning.

**7.1** **Probability** (continued)

**b.** Spin each spinner 20 times (*) and record your results in two tables. Do the data support your answers in part (a)? Explain why or why not?

| Spinner 1 | |
|---|---|
| Number | Frequency |
| 1 | |
| 2 | |
| 3 | |
| 4 | |
| 5 | |
| 6 | |

| Spinner 2 | |
|---|---|
| Number | Frequency |
| 1 | |
| 2 | |
| 3 | |
| 4 | |
| 5 | |
| 6 | |

**c.** How can you use percents to describe the likelihood of spinning each number? Explain.

(*) Spinners are available in the back of the Student Journal.

**Notetaking with Vocabulary**

**Vocabulary:**

**Notes:**

**7.1** **Self-Assessment**

**Use the scale below to rate your understanding of the learning target and the success criteria.**

| 1 | 2 | 3 | 4 |
|---|---|---|---|
| I do not understand. | I can do it with help. | I can do it on my own. | I can teach someone else. |

| | Rating | Date |
|---|:---:|---|
| **7.1 Probability** | | |
| **Learning Target:** Understand how the probability of an event indicates its likelihood. | 1  2  3  4 | |
| I can identify possible outcomes of an experiment. | 1  2  3  4 | |
| I can use probability and relative frequency to describe the likelihood of an event. | 1  2  3  4 | |
| I can use relative frequency to make predictions. | 1  2  3  4 | |

## 7.1 Practice

**You randomly choose one shape from the bag.**
**(a) Find the number of ways the event can occur.**
**(b) Find the favorable outcomes of the event.**

**1.** Choosing a triangle

**2.** Choosing a star

**3.** Choosing *not* a square

**4.** Choosing *not* a circle

**5.** There are 12 cats and 6 dogs at the Humane Society. You randomly choose a pet.

   **a.** How many possible outcomes are there?

   **b.** In how many ways can you choose a cat?

   **c.** In how many ways can you choose a dog?

   **d.** Describe the likelihood that you choose a dog? Explain your reasoning.

**Describe the likelihood of the event from the given information.**

**6.** The school bus arrives late $\frac{2}{7}$ of the time. The school bus arrives on time.

**7.** The probability that it rains during a hurricane is 1. It rains during a hurricane.

**8.** There is an 85% chance that you will go to the concert. You do not go to the concert.

**9.** A store has 30 blue pens, 18 black pens, and 12 red pens in stock. You buy 30 blue pens and 5 red pens. Describe the likelihood of each event before and after your purchase.

   **a.** Randomly choosing a blue pen at the store.

   **b.** Randomly choosing a black pen at the store.

   **c.** Randomly choosing a red pen at the store.

# 7.2 Experimental and Theoretical Probability
### For use with Exploration 7.2

**Learning Target:** Develop probability models using experimental and theoretical probability.

**Success Criteria:**
- I can explain the meanings of experimental probability and theoretical probability.
- I can find experimental and theoretical probabilities.
- I can use probability to make predictions.

---

**1 EXPLORATION: Conducting Experiments**

**Work with a partner. Conduct the following experiments and find the relative frequencies.**

Experiment 1

- Flip a quarter 25 times and record whether each flip lands heads up or tails up.

Experiment 2

- Toss a thumbtack onto a table 25 times and record whether each toss lands point up or on its side.

**7.2** **Experimental and Theoretical Probability** (continued)

**a.** Combine your results with those of your classmates. Do the relative frequencies change? What do you notice?

**b.** Everyone in your school conducts each experiment and you combine the results. How do you expect the relative frequencies to change? Explain.

**c.** How many times in 1000 flips do you expect a quarter to land heads up? How many times in 1000 tosses to you expect a thumbtack to land point up? Explain your reasoning.

**d.** In a *uniform probability model*, each outcome is equally likely to occur. Can you use a uniform probability model to describe either experiment? Explain.

**Notetaking with Vocabulary**

**Vocabulary:**

**Notes:**

**7.2** **Self-Assessment**

Use the scale below to rate your understanding of the learning target and
the success criteria.

| **1** | **2** | **3** | **4** |
|---|---|---|---|
| I do not understand. | I can do it with help. | I can do it on my own. | I can teach someone else. |

|  | Rating | Date |
|---|---|---|
| **7.2 Experimental and Theoretical Probability** | | |
| **Learning Target:** Develop probability models using experimental and theoretical probability. | 1   2   3   4 | |
| I can explain the meanings of experimental probability and theoretical probability. | 1   2   3   4 | |
| I can find experimental and theoretical probabilities. | 1   2   3   4 | |
| I can use probability to make predictions. | 1   2   3   4 | |

## 7.2 Practice

You have four sticks. Two sticks have one blue side and one pink side. One stick has 2 blue sides. One stick has 2 pink sides. You throw the sticks 20 times and record the results. Use the table to find the experimental probability of the event.

| Outcome | Frequency |
|---|---|
| 3 blue, 1 pink | 7 |
| 2 blue, 2 pink | 9 |
| 1 blue, 3 pink | 4 |

1. tossing 1 pink and 3 blue

2. tossing the same number of blue and pink

3. *not* tossing 3 pink

4. tossing at most 2 blue

5. You check 30 containers of yogurt. Seven of them have an expiration date within the next 3 days.

   a. What is the experimental probability that a container of yogurt will have an expiration date within the next 3 days?

   b. Out of 120 containers of yogurt, how many would you expect to have an expiration date within the next 3 days? Explain.

6. You flip 3 coins 50 times, and flipping 3 tails occurs 6 times, flipping 3 heads occurs 7 times.

   a. What is the theoretical probability that you flip 3 heads?

   b. What is the theoretical probability that you flip less than 3 heads?

   c. What is the experimental probability that you flip 3 heads?

   d. What is the experimental probability that you flip 3 tails?

   e. How many times would you expect to flip 3 tails out of 200 trials of flipping 3 coins? Explain.

7. Each letter of the alphabet is printed on an index card. After removing the cards that spell the word "vowel," what is the theoretical probability of randomly choosing a vowel? Explain.

8. There are 12 cats and 8 dogs in the pet store.

   a. What is the theoretical probability that a randomly chosen pet is a cat?

   b. Two days later, there are half as many cats in the pet store. The theoretical probability that a randomly chosen pet is a cat is half that of two days earlier. How many dogs are in the pet store two days later?

## 7.3 Compound Events
**For use with Exploration 7.3**

**Learning Target:** Find sample spaces and probabilities of compound events.

**Success Criteria:**
- I can find the sample space of two or more events.
- I can find the total number of possible outcomes of two or more events.
- I can find probabilities of compound events.

---

**1 EXPLORATION: Comparing Combination Locks**

**Work with a partner. You are buying a combination lock. You have three choices.**

a. One lock has 3 wheels. Each wheel is numbered from 0 to 9. How many possible outcomes are there for each wheel? How many possible combinations are there?

b. How can you use the number of possible outcomes on each wheel to determine the number of possible combinations?

**7.3** Compound Events (continued)

c. Another lock has one wheel numbered from 0 to 39. Each combination uses a sequence of three numbers. How many possible combinations are there?

d. Another lock has 4 wheels as described. How many possible combinations are there?

**Wheel 1:** 0–9

**Wheel 2:** A–J

**Wheel 3:** K–T

**Wheel 4:** 0–9

e. For which lock are you least likely to guess the combination? Why?

**Notetaking with Vocabulary**

**Vocabulary:**

**Notes:**

**7.3** **Self-Assessment**

Use the scale below to rate your understanding of the learning target and the success criteria.

| 1 | 2 | 3 | 4 |
|---|---|---|---|
| I do not understand. | I can do it with help. | I can do it on my own. | I can teach someone else. |

|  | Rating | Date |
|---|---|---|
| **7.3 Compound Events** | | |
| **Learning Target:** Find sample spaces and probabilities of compound events. | 1  2  3  4 | |
| I can find the sample space of two or more events. | 1  2  3  4 | |
| I can find the total number of possible outcomes of two or more events. | 1  2  3  4 | |
| I can find probabilities of compound events. | 1  2  3  4 | |

## 7.3 Practice

**Use a tree diagram to find the sample space and the total number of possible outcomes.**

1.

| Game | |
|---|---|
| **Coin** | Quarter, Dime, Nickel, Penny |
| **Card** | King, Queen, Jack |

**Use the Fundamental Counting Principle to find the total number of possible outcomes.**

2.

| Sandwich | |
|---|---|
| **Bread** | Italian, Wheat |
| **Meat** | Ham, Roast Beef, Salami |
| **Cheese** | American, Provolone, Swiss |

3. Your computer password contains six characters. Each character is either a number or a letter. How many different passwords are possible? If numbers and letters cannot be used more than once, how many passwords are possible?

4. To win the jackpot you must match five numbers from 1 to 40. What is the probability that you will win the jackpot?

5. You need to hang seven pictures in a straight line.

   **a.** In how many ways can this be accomplished?

   **b.** If the picture of your great-grandfather must be in the middle, how many ways can the seven pictures be hung? Explain.

6. A license plate must contain two letters followed by four digits. How many license plates are possible? If the rule changed to five digits instead of four digits, how many more license plates would be possible?

## 7.4  Simulations
### For use with Exploration 7.4

**Learning Target:**  Design and use simulations to find probabilities of compound events.

**Success Criteria:**
- I can design a simulation to model a real-life situation.
- I can recognize favorable outcomes in a simulation.
- I can use simulations to find experimental probabilities.

---

**1  EXPLORATION: Using a Simulation**

**Work with a partner. A basketball player makes 80% of her free throw attempts.**

a. Is she likely to make at least two of her next three free throws? Explain your reasoning.

b. The table shows 30 randomly generated numbers from 0 to 999. Let each number represent three shots. How can you use the digits of these numbers to represent made shots and missed shots?

| 838 | 617 | 282 | 341 | 785 |
|-----|-----|-----|-----|-----|
| 747 | 332 | 279 | 082 | 716 |
| 937 | 308 | 800 | 994 | 689 |
| 198 | 025 | 853 | 591 | 813 |
| 672 | 289 | 518 | 649 | 540 |
| 865 | 631 | 227 | 004 | 840 |

**7.4** **Simulations** (continued)

c. Use the table to estimate the probability that of her next three free throws, she makes

- exactly two free throws.

- at most one free throw.

- at least two free throws.

- at least two free throws in a row.

d. The experiment used in parts (b) and (c) is called a *simulation*. Another player makes $\frac{3}{5}$ of her free throws. Describe a simulation that can be used to estimate the probability that she makes three of her next four free throws.

##  7.4 Notetaking with Vocabulary

**Vocabulary:**

**Notes:**

## 7.4 Self-Assessment

Use the scale below to rate your understanding of the learning target and the success criteria.

| **1** | **2** | **3** | **4** |
|---|---|---|---|
| I do not understand. | I can do it with help. | I can do it on my own. | I can teach someone else. |

|  | Rating | Date |
|---|---|---|
| **7.4 Simulations** | | |
| **Learning Target:** Design and use simulations to find probabilities of compound events. | 1  2  3  4 | |
| I can design a simulation to model a real-life situation. | 1  2  3  4 | |
| I can recognize favorable outcomes in a simulation. | 1  2  3  4 | |
| I can use simulations to find experimental probabilities. | 1  2  3  4 | |

Name _____ Date _____

## 7.4 Practice

A medicine is effective on 70% of patients. The table shows 30 randomly generated numbers from 0 to 999. Use the table to estimate the probability of the event.

| 028 | 837 | 618 | 205 | 984 |
|-----|-----|-----|-----|-----|
| 724 | 301 | 249 | 946 | 925 |
| 042 | 113 | 696 | 985 | 632 |
| 312 | 085 | 997 | 198 | 398 |
| 117 | 240 | 853 | 373 | 597 |
| 606 | 077 | 016 | 012 | 695 |

1. The medicine is effective on at least two of the next three patients.

2. The medicine is effective on none of the next three patients.

**Design and use a simulation to find the experimental probability.**

3. A bowler hits the head pin 90% of the time that all ten pins are standing. What is the experimental probability that the bowler hits the head pin exactly four of the next five times that all ten pins are standing?

4. In your garden, 40% of your pineapple plants produce a pineapple each year. What is the experimental probability that at most two of the three pineapple plants produce a pineapple this year? Explain.

5. A fitness center randomly selects one of five free gifts to send to each new customer. Gifts include a workout towel, a water bottle, a heart monitor, a T-shirt, and a gift card to the marketplace. What is the experimental probability that for the next two customers, one receives a heart monitor and one receives a gift card to the marketplace? Explain.

**Design and use a simulation with number cubes to estimate the probability.**

6. Your younger sibling will finish his or her dinner plate two out of every six dinners. Estimate the probability that at least two of the next three dinner plates are finished. Explain.

7. You win your favorite video game $\frac{5}{6}$ of the time. Estimate the probability that you will win your favorite video game exactly three of the next four times you play the game. Explain.

Name_____ Date_____

# Chapter Self-Assessment

Use the scale below to rate your understanding of the learning target and the success criteria.

| **1** | **2** | **3** | **4** |
|---|---|---|---|
| I do not understand. | I can do it with help. | I can do it on my own. | I can teach someone else. |

|  | Rating | Date |
|---|---|---|
| **7.1 Probability** | | |
| **Learning Target:** Understand how the probability of an event indicates its likelihood. | 1  2  3  4 | |
| I can identify possible outcomes of an experiment. | 1  2  3  4 | |
| I can use probability and relative frequency to describe the likelihood of an event. | 1  2  3  4 | |
| I can use relative frequency to make predictions. | 1  2  3  4 | |
| **7.2 Experimental and Theoretical Probability** | | |
| **Learning Target:** Develop probability models using experimental and theoretical probability. | 1  2  3  4 | |
| I can explain the meanings of experimental probability and theoretical probability. | 1  2  3  4 | |
| I can find experimental and theoretical probabilities. | 1  2  3  4 | |
| I can use probability to make predictions. | 1  2  3  4 | |
| **7.3 Compound Events** | | |
| **Learning Target:** Find sample spaces and probabilities of compound events. | 1  2  3  4 | |
| I can find the sample space of two or more events. | 1  2  3  4 | |
| I can find the total number of possible outcomes of two or more events. | 1  2  3  4 | |
| I can find probabilities of compound events. | 1  2  3  4 | |

# Chapter 7 Chapter Self-Assessment (continued)

| | Rating | | | | Date |
|---|---|---|---|---|---|
| **7.4 Simulations** | | | | | |
| **Learning Target:** Design and use simulations to find probabilities of compound events. | 1 | 2 | 3 | 4 | |
| I can design a simulation to model a real-life situation. | 1 | 2 | 3 | 4 | |
| I can recognize favorable outcomes in a simulation. | 1 | 2 | 3 | 4 | |
| I can use simulations to find experimental probabilities. | 1 | 2 | 3 | 4 | |

# Review & Refresh

**Find the mean of the data.**

**1.** 15, 21, 10, 18, 26, 23, 18, 20, 21, 19, 25, 22

**2.** 76, 105, 98, 198, 103, 84, 112, 88, 101, 81, 92

**3.** 47.5, 47.3, 47.2, 47.4, 47.9, 47.5, 47.1, 47.6, 47.7

**4.** Identify any outliers in Exercises 1–3. Describe the effect of the outlier on the mean.

**Big Ideas Math: Modeling Real Life Grade 7 Accelerated** **177**
Student Journal

**Find the median, first quartile, third quartile, and interquartile range of the data.**

5. 45, 68, 59, 38, 21, 26, 58, 46, 33, 41, 52, 51, 25

6. 187, 225, 174, 179, 231, 208, 110, 200, 194, 215, 199, 212

7. 2.6, 4.9, 1.4, 1.7, 2.3, 9.4, 3.8, 3.5, 7.6, 4.4, 5.0, 3.1, 2.6, 6.8, 9.1

8. Use the interquartile range to identify any outliers in Exercises 5–7.

# 8.1 Samples and Populations
**For use with Exploration 8.1**

**Learning Target:** Understand how to use random samples to make conclusions about a population.

**Success Criteria:**
- I can explain why a sample is biased or unbiased.
- I can explain why conclusions made from a biased sample may not be valid.
- I can use an unbiased sample to make a conclusion about a population.

---

 **EXPLORATION:** Using Samples of Populations

**Work with a partner. You want to make conclusions about the favorite extracurricular activities of students at your school.**

a. Identify the population. Then identify five samples of the population.

b. When a sample is selected *at random*, each member of the population is equally likely to be selected. Are any of the samples in part (a) selected at random? Explain your reasoning.

**Big Ideas Math: Modeling Real Life Grade 7 Accelerated** **179**
Student Journal

**8.1**  **Samples and Populations** (continued)

c. How are the samples below different? Is each conclusion valid? Explain your reasoning.

> You ask 20 members of the school band about their favorite activity. The diagram shows the results. You conclude that band is the favorite activity of 70% of the students in your school.

**Favorite Activity**

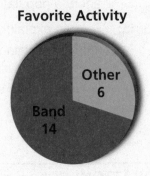

Other 6

Band 14

> You ask every eighth student who enters the school about their favorite activity. One student says glee club for every nine that name a different activity. You conclude that glee club is the favorite activity of 10% of the students in your school.

d. Write a survey question about a topic that interests you. How can you choose people to survey so that you can use the results to make a valid conclusion?

## 8.1 Notetaking with Vocabulary

**Vocabulary:**

**Notes:**

## 8.1 Self-Assessment

**Use the scale below to rate your understanding of the learning target and the success criteria.**

| *1* | *2* | *3* | *4* |
|---|---|---|---|
| I do not understand. | I can do it with help. | I can do it on my own. | I can teach someone else. |

|  | Rating | Date |
|---|---|---|
| **8.1 Samples and Populations** | | |
| **Learning Target:** Understand how to use random samples to make conclusions about a population. | 1  2  3  4 | |
| I can explain why a sample is biased or unbiased. | 1  2  3  4 | |
| I can explain why conclusions made from a biased sample may not be valid. | 1  2  3  4 | |
| I can use an unbiased sample to make a conclusion about a population. | 1  2  3  4 | |

# 8.1 Practice

1. You want to know the number of fans at the Miami Dolphins and Dallas Cowboys game that think the Dolphins will win. You survey 50 fans with season tickets for the Dolphins.

   a. What is the population of your survey?

   b. What is the sample of your survey?

   c. Is the sample biased or unbiased? Explain.

2. Which sample is better for making an estimate? Explain.

| Estimate the number of families in your town with two or more children. | |
|---|---|
| Sample A | A random sample of 10 families living near your home |
| Sample B | A random sample of 10 families living in your town |

**Determine whether you would survey the population or a sample. Explain.**

3. You want to know the favorite clothing store of the students at your school.

4. You want to know the favorite topic of students in your history class.

5. An administrator surveys a random sample of 48 out of 900 middle school students. Using the survey results, the administrator estimates that 225 students are in favor of the new dress code. How many of the 48 students surveyed were in favor of the new dress code?

6. The table show the results of a survey of 75 randomly chosen seventh graders from your school. There are 300 seventh graders in your school. In the survey, each individual was asked to name his or her favorite type of music.

| Music | Frequency |
|---|---|
| Rock | 20 |
| Country | 23 |
| Rap | 30 |
| Classical | 2 |

   a. Estimate the number of seventh graders in your school whose favorite type of music is country.

   b. If you were to repeat the survey using randomly chosen adults, would you predict that the results of the adult survey will be different if you surveyed adults in their 30s versus adults in their 70s? Explain your reasoning.

# 8.2 Using Random Samples to Describe Populations

**For use with Exploration 8.2**

**Learning Target:** Understand variability in samples of a population.

**Success Criteria:**
- I can use multiple random samples to make conclusions about a population.
- I can use multiple random samples to examine variation in estimates.

---

**1 EXPLORATION: Exploring Variability in Samples**

**Work with a partner. Sixty percent of all seventh graders have visited a planetarium.**

**a.** Design a simulation using packing peanuts. Mark 60% of the packing peanuts and put them in a paper bag. What does choosing a marked peanut represent?

**b.** Simulate a sample of 25 students by choosing peanuts from the bag, replacing the peanut each time. Record the results.

### 8.2 Using Random Samples to Describe Populations (continued)

c. Find the percent of students in the sample who have visited a planetarium. Compare this value to the actual percent of all seventh graders who have visited a planetarium.

d. Record the percent in part (c) from each pair in the class. Use a dot plot to display the data. Describe the variation in the data.

## 8.2 Notetaking with Vocabulary

**Vocabulary:**

**Notes:**

## 8.2 Self-Assessment

Use the scale below to rate your understanding of the learning target and the success criteria.

| **1** | **2** | **3** | **4** |
|---|---|---|---|
| I do not understand. | I can do it with help. | I can do it on my own. | I can teach someone else. |

| | Rating | Date |
|---|---|---|
| **8.2 Using Random Samples to Describe Populations** | | |
| **Learning Target:** Understand variability in samples of a population. | 1  2  3  4 | |
| I can use multiple random samples to make conclusions about a population. | 1  2  3  4 | |
| I can use multiple random samples to examine variation in estimates. | 1  2  3  4 | |

# 8.2 Practice

1. Work with a partner. Mark 30 small pieces of paper with an A, a B, or a C. Put the pieces of paper in a bag. Trade bags with other students in the class.

   a. Generate a sample by choosing a piece of paper from your bag 10 times, replacing the piece of paper each time. Record the number of times you choose each letter. Repeat this process to generate five more samples.

   b. Use each sample to make an estimate for the number of As and Bs in the bag. Then describe the variation of the six estimates. Make estimates for the numbers of As, Bs, and Cs in the bag based on all the samples.

   c. Take the pieces of paper out of the bag. How do your estimates compare to the population? Do you think you can make a more accurate estimate? If so, explain how.

2. Work with a partner. You want to know the mean number of hours students in band or orchestra practice their instruments each week. Prior research indicates that the maximum number of hours of practice is 14 hours per week.

   a. Use technology to simulate the hours of practice for 10 students in band or orchestra. Randomly generate 10 numbers from 0 to 14. Write down the results. Repeat the simulation 9 more times, writing down the results each time.

   b. Use each sample to make an estimate for the mean number of hours students in band or orchestra practice their instruments each week. Describe the variation of the estimates.

   c. Use all ten samples to make one estimate for the mean number of hours students in band or orchestra practice their instruments each week. How does your estimate compare to the mean of the entire data set?

   d. Use each sample to make an estimate for the median number of hours students in band or orchestra practice their instruments each week. Describe the variation of the estimates.

   e. Use all ten samples to make one estimate for the median number of hours students in band or orchestra practice their instruments each week.

Name_____ Date_____

## 8.3 Comparing Populations
**For use with Exploration 8.3**

**Learning Target:** Compare populations using measures of center and variation.

**Success Criteria:**
- I can find the measures of center and variation of a data set.
- I can describe the visual overlap of two data distributions numerically.
- I can determine whether there is a significant difference in the measures of center of two data sets.

---

**1  EXPLORATION: Comparing Two Data Distributions**

**Work with a partner.**

**a.** Does each data display show *overlap*? Explain.

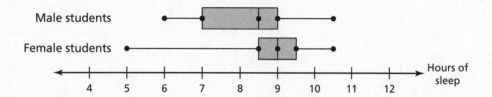

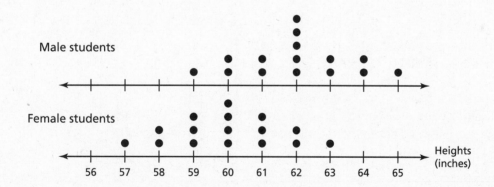

**8.3** **Comparing Populations** (continued)

**Ages of People in Two Exercise Classes**

| 10:00 A.M. Class | | | | | | | | 8:00 P.M. Class | | | | | |
|---|---|---|---|---|---|---|---|---|---|---|---|---|---|
| | | | | | | 1 | 8 | 9 | | | | | |
| | | | | | | 2 | 1 | 2 | 2 | 7 | 9 | 9 | |
| | | | | | | 3 | 0 | 3 | 4 | 5 | 7 | | |
| 9 | 7 | 3 | 2 | 2 | 2 | 4 | 0 | | | | | | |
| | 7 | 5 | 4 | 3 | 1 | 5 | | | | | | | |
| | | 7 | 0 | 0 | 6 | | | | | | | | |
| | | | 0 | 7 | | | | | | | | | |

**Key: 2 | 4 | 0 = 42 and 40 years**

**b.** How can you describe the overlap of two data distributions using words? How can you describe the overlap numerically?

**c.** In which pair of data sets is the difference in the measures of center the most significant? Explain your reasoning.

**Notetaking with Vocabulary**

**Vocabulary:**

**Notes:**

**8.3 Self-Assessment**

Use the scale below to rate your understanding of the learning target and the success criteria.

| 1 | 2 | 3 | 4 |
|---|---|---|---|
| I do not understand. | I can do it with help. | I can do it on my own. | I can teach someone else. |

| | Rating | Date |
|---|---|---|
| **8.3 Comparing Populations** | | |
| **Learning Target:** Compare populations using measures of center and variation. | 1  2  3  4 | |
| I can find the measures of center and variation of a data set. | 1  2  3  4 | |
| I can describe the visual overlap of two data distributions numerically. | 1  2  3  4 | |
| I can determine whether there is a significant difference in the measures of center of two data sets. | 1  2  3  4 | |

## 8.3 Practice

1. A scientist experiments with a new type of corn. The dot plots show the heights of corn stalks in two gardens. Experimental corn was planted in Garden A. Traditional corn was planted in Garden B.

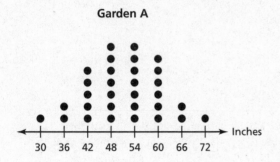

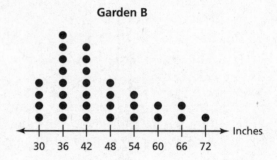

a. Compare the data sets using measures of center and variation.

b. Express the difference in the measures of center as a multiple of the measure of variation.

c. If the experimental corn stalks are significantly taller than the traditional corn, the scientist will repeat the experiment in a field rather than a garden. Should the scientist repeat the experiment in a field? Justify your answer.

2. The double box-and-whisker plot represents the runs scored per game by two softball teams during a 15-game season.

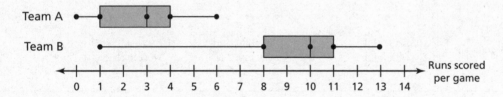

a. Compare the data sets using measures of center and variation.

b. Which data set is more likely to contain a value of 3?

c. Is the number of runs scored per game significantly greater for one team than the other? Explain.

Name_____ Date_____

## 8.4 Using Random Samples to Compare Populations
**For use with Exploration 8.4**

**Learning Target:** Use random samples to compare populations.

**Success Criteria:**
- I can compare random samples using measures of center and variation.
- I can recognize whether random samples are likely to be representative of a population.
- I can compare populations using multiple random samples.

---

**1  EXPLORATION:** Using Random Samples

Work with a partner. You want to compare the numbers of hours spent on homework each week by male and female students in your state. You take a random sample of 15 male students and 15 female students throughout the state.

| Male Students | | | | |
|------|-----|---|-----|---|
| 1.5 | 3 | 0 | 2.5 | 1 |
| 8 | 2.5 | 1 | 3 | 0 |
| 6.5 | 1 | 5 | 0 | 5 |

| Female Students | | | | |
|---|---|---|-----|----|
| 4 | 0 | 3 | 1 | 1 |
| 5 | 1 | 3 | 5.5 | 10 |
| 2 | 0 | 6 | 3.5 | 2 |

**a.** Compare the data in each sample.

**b.** Are the samples likely to be representative of all male and female students in your state? Explain.

---

**Big Ideas Math: Modeling Real Life Grade 7 Accelerated**

**8.4** **Using Random Samples to Compare Populations** (continued)

c. You take 100 random samples of 15 male students in your state and 100 random samples of 15 female students in your state and record the median of each sample. The double box-and-whisker plot shows the distributions of the sample medians. Compare the distributions in the double box-and-whisker plot with the distributions of the data in the tables.

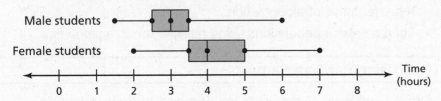

d. What can you conclude from the double box-and-whisker plot? Explain.

e. How can you use random samples to make accurate comparisons of two populations?

Name_____ Date_____

## 8.4 Notetaking with Vocabulary

**Vocabulary:**

**Notes:**

## 8.4 Self-Assessment

**Use the scale below to rate your understanding of the learning target and the success criteria.**

1 I do not understand. 2 I can do it with help. 3 I can do it on my own. 4 I can teach someone else.

| | Rating | Date |
|---|---|---|
| **8.4 Using Random Samples to Compare Populations** | | |
| **Learning Target:** Use random samples to compare populations. | 1 2 3 4 | |
| I can compare random samples using measures of center and variation. | 1 2 3 4 | |
| I can recognize whether random samples are likely to be representative of a population. | 1 2 3 4 | |
| I can compare populations using multiple random samples. | 1 2 3 4 | |

## 8.4 Practice

1. The tables show the numbers of attendees at random pep rallies for football and basketball games at a school during the last 5 years.

**Football Pep Rally Attendance**

| 174 | 175 | 200 | 169 | 178 | 171 |
|-----|-----|-----|-----|-----|-----|
| 165 | 187 | 159 | 170 | 184 | 196 |
| 205 | 231 | 198 | 310 | 152 | 178 |

**Basketball Pep Rally Attendance**

| 143 | 178 | 154 | 167 | 204 | 199 |
|-----|-----|-----|-----|-----|-----|
| 254 | 147 | 179 | 162 | 189 | 203 |
| 217 | 214 | 187 | 210 | 288 | 287 |

   a. Find the mean, median, mode, range, interquartile range, and mean absolute deviation for the samples.

   b. When comparing the two samples using measures of center and variation, would you use the mean and the MAD, or the median and the IQR? Explain.

   c. Compare the samples.

   d. Can you determine which pep rally is more popular?

2. Two manufacturing plants each produce 120 frozen food products. A scientist takes 10 samples of 12 frozen food products from each lab and records the number that pass an inspection. Are the samples likely to be representative of all the frozen food products for each manufacturing plant? If so, which lab has more frozen food products that will pass the inspection? Justify your answer.

**Manufacturing Plant A**

| 9 | 8 | 12 | 7 | 9 |
|---|---|----|---|---|
| 10 | 7 | 11 | 8 | 9 |

**Manufacturing Plant B**

| 12 | 11 | 11 | 10 | 12 |
|----|----|----|----|----|
| 12 | 10 | 11 | 12 | 12 |

Name_____ Date _____

# Chapter Self-Assessment

Use the scale below to rate your understanding of the learning target and the success criteria.

| **1** | **2** | **3** | **4** |
|---|---|---|---|
| I do not understand. | I can do it with help. | I can do it on my own. | I can teach someone else. |

|  | Rating | Date |
|---|---|---|
| **8.1 Samples and Populations** | | |
| **Learning Target:** Understand how to use random samples to make conclusions about a population. | 1  2  3  4 | |
| I can explain why a sample is biased or unbiased. | 1  2  3  4 | |
| I can explain why conclusions made from a biased sample may not be valid. | 1  2  3  4 | |
| I can use an unbiased sample to make a conclusion about a population. | 1  2  3  4 | |
| **8.2 Using Random Samples to Describe Populations** | | |
| **Learning Target:** Understand variability in samples of a population. | 1  2  3  4 | |
| I can use multiple random samples to make conclusions about a population. | 1  2  3  4 | |
| I can use multiple random samples to examine variation in estimates. | 1  2  3  4 | |
| **8.3 Comparing Populations** | | |
| **Learning Target:** Compare populations using measures of center and variation. | 1  2  3  4 | |
| I can find the measures of center and variation of a data set. | 1  2  3  4 | |
| I can describe the visual overlap of two data distributions numerically. | 1  2  3  4 | |
| I can determine whether there is a significant difference in the measures of center of two data sets. | 1  2  3  4 | |

# Chapter Self-Assessment (continued)

| | Rating | Date |
|---|---|---|
| **8.4 Using Random Samples to Compare Populations** | | |
| **Learning Target:** Use random samples to compare populations. | 1　2　3　4 | |
| I can compare random samples using measures of center and variation. | 1　2　3　4 | |
| I can recognize whether random samples are likely to be representative of a population. | 1　2　3　4 | |
| I can compare populations using multiple random samples. | 1　2　3　4 | |

## Chapter 9 Review & Refresh

**Identify the basic shapes in the figure.**

1.

2.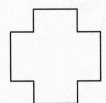

3.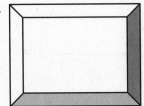

4.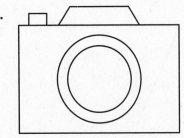

5. Identify the basic shapes that make up the top of the teacher's desk.

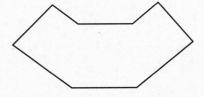

**Chapter 9** **Review & Refresh** (continued)

**Use a protractor to find the measure of the angle. Then classify the angle as *acute*, *obtuse*, *right*, or *straight*.**

**6.**

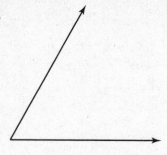

**7.**

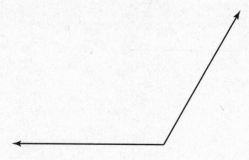

**8.**

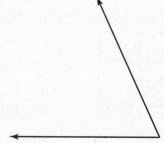

**9.**

**Use a protractor to draw an angle with the given measure.**

**10.** 80°

**11.** 175°

# 9.1 Circles and Circumference
### For use with Exploration 9.1

**Learning Target:** Find the circumference of a circle.

**Success Criteria:**
- I can explain the relationship between the diameter and circumference of a circle.
- I can use a formula to find the circumference of a circle.

---

**1 EXPLORATION: Using a Compass to Draw a Circle**

**Work with a partner. Set a compass to 2 inches and draw a circle.**

a. Draw a line from one side of the circle to the other that passes through the center. What is the length of the line? This is called the *diameter* of the circle.

b. Estimate the distance around the circle. This is called the *circumference* of the circle. Explain how you found your answer.

**9.1** **Circles and Circumference** (continued)

**2** **EXPLORATION:** Exploring Diameter and Circumference

**Work with a partner.**

a. Roll a cylindrical object on a flat surface to find the circumference of the circular base.

b. Measure the diameter of the circular base. Which is greater, the diameter or the circumference? how many times greater?

c. Compare your answers in part (b) with the rest of the class. What do you notice?

d. Without measuring, how can you find the circumference of a circle with a given diameter? Use your method to estimate the circumference of the circle in Exploration 1.

# 9.1 Notetaking with Vocabulary

**Vocabulary:**

**Notes:**

# 9.1 Self-Assessment

Use the scale below to rate your understanding of the learning target and
the success criteria.

| *1* | *2* | *3* | *4* |
|---|---|---|---|
| I do not understand. | I can do it with help. | I can do it on my own. | I can teach someone else. |

|  | Rating | Date |
|---|---|---|
| **9.1 Circles and Circumference** | | |
| **Learning Target:** Find the circumference of a circle. | 1  2  3  4 | |
| I can explain the relationship between the diameter and circumference of a circle. | 1  2  3  4 | |
| I can use a formula to find the circumference of a circle. | 1  2  3  4 | |

# 9.1 Practice

**Find the circumference of the circle. Use 3.14 or $\frac{22}{7}$ for $\pi$.**

1.
26 m

2.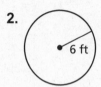
6 ft

3.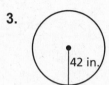
42 in.

**Find the perimeter of the semicircular region.**

4.
28 yd

5.
7.5 ft

6.
11 cm

7. Consider the circles $A$, $B$, $C$, and $D$.

A
2.5 ft

B
24 in.

C
32 ft

D
84 in.

a. Without calculating, which circle has the greatest circumference? Explain.

b. Without calculating, which circle has the least circumference? Explain.

c. Find the circumference of each circle in feet. How many times larger is the greatest circumference than the least circumference? How could you find this answer without calculating?

8. A coaster has a circumference of 12.56 inches. Suppose the same amount of coaster is visible around the bottom of a glass as shown. What is the circumference of the glass?

0.5 in.    0.5 in.

9. Are the side lengths of the squares in Diagram A and Diagram B equivalent? Explain your reasoning.

Diagram A    Diagram B

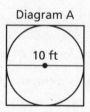
10 ft

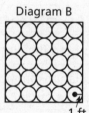
1 ft

10. You release a ball with a radius of 1 inch into a pipe as shown. How many times will the ball rotate before it falls out of the other end of the pipe?

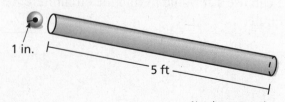
1 in.

5 ft

*Not drawn to scale*

Name_____ Date_____

## 9.2 Areas of Circles
### For use with Exploration 9.2

**Learning Target:** Find the area of a circle.

**Success Criteria:**
- I can estimate the area of a circle.
- I can use a formula to find the area of a circle.

### 1 EXPLORATION: Estimating the Area of a Circle

Work with a partner. Each grid contains a circle with a diameter of 4 centimeters. Use each grid to estimate the area of the circle. Which estimate should be closest to the actual area? Explain.

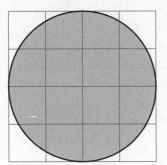 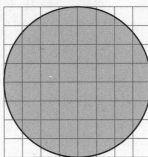 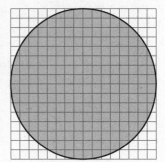

Name _____  Date _____

---

**2** **EXPLORATION:** Writing a Formula for the Area of a Circle

**Work with a partner. A student draws a circle with radius $r$ and divides the circle into 24 equal sections (\*). The student cuts out each section and arranges the sections to form a shape that resembles a parallelogram.**

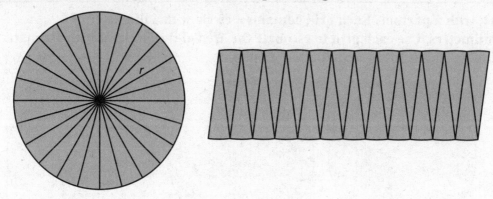

a.  Use the diagram to write a formula for the area $A$ of a circle in terms of the radius $r$. Explain your reasoning.

b.  Use the formula to check your estimates in Exploration 1.

\* Circle is available in the back of the Student Journal.

**Notetaking with Vocabulary**

**Vocabulary:**

**Notes:**

**9.2** **Self-Assessment**

Use the scale below to rate your understanding of the learning target and the success criteria.

| *1* | *2* | *3* | *4* |
|---|---|---|---|
| I do not understand. | I can do it with help. | I can do it on my own. | I can teach someone else. |

|  | Rating | Date |
|---|---|---|
| **9.2 Areas of Circles** | | |
| **Learning Target:** Find the area of a circle. | 1  2  3  4 | |
| I can estimate the area of a circle. | 1  2  3  4 | |
| I can use a formula to find the area of a circle. | 1  2  3  4 | |

Name _____ Date _____

## 9.2 Practice

**Find the area of the circle. Use 3.14 or $\frac{22}{7}$ for $\pi$.**

1.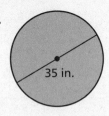

35 in.

2.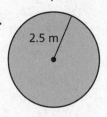

2.5 m

3.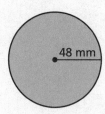

48 mm

**Find the area of the semicircle.**

4.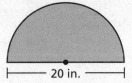

20 in.

5.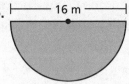

16 m

6.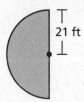

21 ft

7. The shadow of an object is roughly the same size as the object. What is the area of the circular shadow of the hot air balloon?

56 ft

8. How many *square feet* of the ground are sprayed by the beach shower?

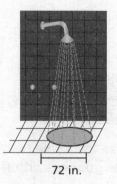

72 in.

9. The radius of the small circle is half the radius of the large circle.

a. Use the radius $r$ to write a formula for the area of the large circle.

b. Use the radius $\frac{r}{2}$ to write a formula for the area of the small circle.

c. How does the area of a circle compare to the area of another circle whose radius is twice as large? Explain your reasoning.

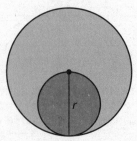

10. The number of square inches of a circle's area is equal to the number of inches of its circumference. What is the radius of the circle? Explain how you found your answer.

# 9.3 Perimeters and Areas of Composite Figures
### For use with Exploration 9.3

**Learning Target:** Find perimeters and areas of composite figures.

**Success Criteria:**
- I can use a grid to estimate perimeters and areas.
- I can identify the shapes that make up a composite figure.
- I can find the perimeters and areas of shapes that make up composite figures.

---

**1  EXPLORATION: Submitting a Bid**

Work with a partner. You want to bid on a project for the pool shown. The project involves ordering and installing the brown tile that borders the pool, and ordering a custom-made tarp to cover the surface of the pool. In the figure, each grid square represents 1 square foot.

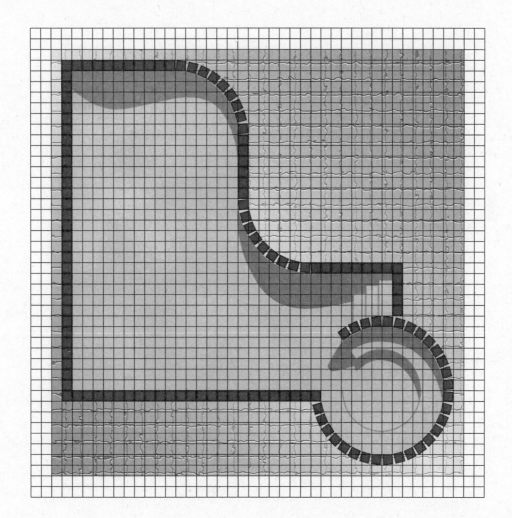

---

## 9.3 Perimeters and Areas of Composite Figures (continued)

- **You pay $5 per linear foot for the tile.**

- **You pay $4 per square foot for the tarp.**

- **It takes you about 15 minutes to install each foot of tile.**

**a.** Estimate the total cost for the tile and the tarp.

**b.** Write a bid for how much you will charge for the project. Include the hourly wage you will receive. Estimate your total profit.

## Notetaking with Vocabulary

**Vocabulary:**

**Notes:**

## 9.3 Self-Assessment

Use the scale below to rate your understanding of the learning target and the success criteria.

| **1** | **2** | **3** | **4** |
|---|---|---|---|
| I do not understand. | I can do it with help. | I can do it on my own. | I can teach someone else. |

|  | Rating | Date |
|---|---|---|
| **9.3 Perimeters and Areas of Composite Figures** | | |
| **Learning Target:** Find perimeters and areas of composite figures. | 1   2   3   4 | |
| I can use a grid to estimate perimeters and areas. | 1   2   3   4 | |
| I can identify the shapes that make up a composite figure. | 1   2   3   4 | |
| I can find the perimeters and areas of shapes that make up composite figures. | 1   2   3   4 | |

Name _____ Date _____

**Estimate the perimeter and the area of the shaded figure.**

**1.**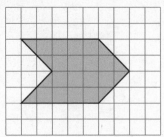

**2.**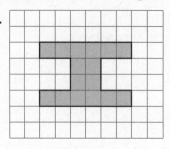

**3.**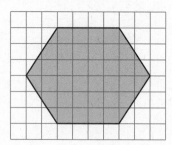

**Find the perimeter and the area of the figure.**

**4.**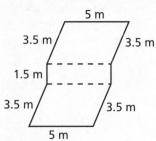

**5.**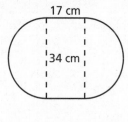

**6.**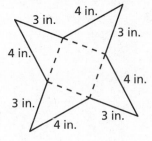

**7.** Describe and correct the error in finding the perimeter of the figure.

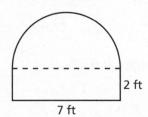

$\times$ Perimeter $\approx 2 + 7 + 2 + 21.98$
$= 32.98$ ft

**8.** A shrub has been cut and trimmed into the shape of an "F". The owner has hired a landscaper to decrease the perimeter of the shrub by 10 feet. Draw a diagram of how the landscaper might do this. Is there more than one way? Explain.

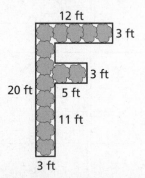

**9.** A school has a garden in the shape of a pencil. A fence is to be built around the garden. The fence costs $2.75 per foot. How much will it cost to install the fence?

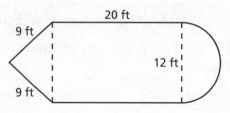

## 9.4 Constructing Polygons
**For use with Exploration 9.4**

**Learning Target:** Construct a polygon with given measures.

**Success Criteria:**
- I can use technology to draw polygons.
- I can determine whether given measures result in one triangle, many triangles, or no triangle.
- I can draw polygons given angle measures or side lengths.

---

**1  EXPLORATION:** Using Technology to Draw Polygons

**Work with a partner.**

a. Use geometry software to draw each polygon with the given side lengths or angle measures, if possible. Complete the table.

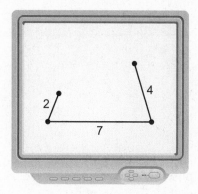

| Side Lengths or Angle Measures | How many figures are possible? |
|---|---|
| i. 4 cm, 6 cm, 7 cm | |
| ii. 2 cm, 6 cm, 7 cm | |
| iii. 2 cm, 4 cm, 7 cm | |
| iv. 2 cm, 4 cm, 6 cm | |
| v. 2 in., 3 in., 3 in., 6 in. | |
| vi. 1 in., 1 in., 3 in., 6 in. | |
| vii. 1 in., 1 in., 3 in., 4 in. | |
| viii. 90°, 60°, 30° | |
| ix. 100°, 40°, 20° | |
| x. 50°, 60°, 70° | |
| xi. 20°, 80°, 100° | |
| xii. 20°, 50°, 50°, 60° | |
| xiii. 30°, 80°, 120°, 130° | |
| xiv. 60°, 60°, 120°, 120° | |

**9.4** **Constructing Polygons** (continued)

**b.** Without constructing, how can you tell whether it is possible to draw a triangle given three angle measures? three side lengths? Explain your reasoning.

**c.** Without constructing, how can you tell whether it is possible to draw a quadrilateral given four angle measures? four side lengths? Explain your reasoning.

## 9.4 Notetaking with Vocabulary

**Vocabulary:**

**Notes:**

## 9.4 Self-Assessment

**Use the scale below to rate your understanding of the learning target and the success criteria.**

| 1 | 2 | 3 | 4 |
|---|---|---|---|
| I do not understand. | I can do it with help. | I can do it on my own. | I can teach someone else. |

|  | Rating | Date |
|---|---|---|
| **9.4 Constructing Polygons** | | |
| **Learning Target:** Construct a polygon with given measures. | 1  2  3  4 | |
| I can use technology to draw polygons. | 1  2  3  4 | |
| I can determine whether given measures result in one triangle, many triangles, or no triangle. | 1  2  3  4 | |
| I can draw polygons given angle measures or side lengths. | 1  2  3  4 | |

## 9.4 Practice

**Draw a triangle with the given angle measures, if possible.**

1. $25°, 65°, 90°$

2. $45°, 60°, 75°$

3. Draw a triangle with the given description: a $110°$ angle connects to a $25°$ angle by a side of length 6 inches.

**Determine whether you can construct *one*, *many*, or *no* triangle(s) with the given description. Explain your reasoning.**

4. a triangle with a side length of 2 inches, a side length of 4 inches, and a side length of 5 inches

5. a scalene triangle with two side lengths of 7 centimeters

6. a triangle with one angle measure of $100°$ and one side length of 6 inches

7. Determine whether the statement is *true* or *false*. Explain your reasoning. You may use diagrams to explain your reasoning.

   a. A rectangle with side lengths of 30 inches and 10 inches can be divided into two congruent squares.

   b. A rectangle with side lengths of 30 inches and 10 inches can be divided into three congruent squares.

   c. A parallelogram with opposite congruent side lengths of 6 feet and 3 feet can be divided into two congruent rhombuses.

   d. A rectangle with side lengths of 30 inches and 10 inches can be divided into two congruent trapezoids.

   e. A rhombus that has side lengths of 8 meters can be divided into two congruent parallelograms.

**Construct a quadrilateral with the given description.**

8. a kite with side lengths of 2 feet and 5 feet

9. a trapezoid with a base of length 4 meters and an opposite parallel side of length 1 meter

10. a rhombus with adjacent angles of $70°$ and $110°$.

## 9.5 Finding Unknown Angle Measures
**For use with Exploration 9.5**

**Learning Target:** Use facts about angle relationships to find unknown angle measures.

**Success Criteria:**
- I can identify adjacent, complementary, supplementary, and vertical angles.
- I can use equations to find unknown angle measures.
- I can find unknown angle measures in real-life situations.

---

**1 EXPLORATION: Using Rules About Angles**

**Work with a partner. The diagram shows pairs of *adjacent* angles and *vertical* angles. Vertical angles cannot be adjacent.**

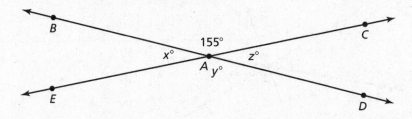

**a.** Which pair(s) of angles are adjacent angles? Explain.

**b.** Which pair(s) of angles are vertical angles? Explain.

**c.** Without using a protractor, find the values of $x$, $y$, and $z$. Explain your reasoning.

---

**9.5** **Finding Unknown Angle Measures** (continued)

**d.** Make a conjecture about the measures of any two vertical angles.

**e.** Test your conjecture in part (d) using the diagram below. Explain why your conjecture is or is *not* true.

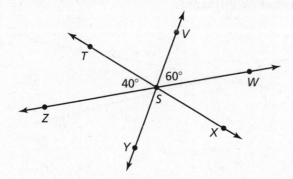

# 9.5 Notetaking with Vocabulary

**Vocabulary:**

**Notes:**

# 9.5 Self-Assessment

Use the scale below to rate your understanding of the learning target and the success criteria.

| *1* | *2* | *3* | *4* |
|---|---|---|---|
| I do not understand. | I can do it with help. | I can do it on my own. | I can teach someone else. |

|  | Rating | Date |
|---|---|---|
| **9.5 Finding Unknown Angle Measures** | | |
| **Learning Target:** Use facts about angle relationships to find unknown angle measures. | 1  2  3  4 | |
| I can identify adjacent, complementary, supplementary, and vertical angles. | 1  2  3  4 | |
| I can use equations to find unknown angle measures. | 1  2  3  4 | |
| I can find unknown angle measures in real-life situations. | 1  2  3  4 | |

## 9.5  Practice

**Name two pairs of adjacent angles and two pairs of vertical angles in the figure.**

1.
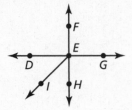

2.
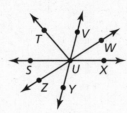

**Classify the pair of angles. Then find the value of x.**

3.
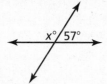

4.

5.
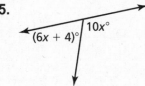

6.

7. Draw a figure in which $\angle 2$ and $\angle 3$ are acute vertical angles, $\angle 1$ and $\angle 2$ are supplementary angles, $\angle 2$ and $\angle 5$ are complementary angles, and $\angle 4$ and $\angle 5$ are adjacent angles.

8. Find the values of $x$ and $y$.

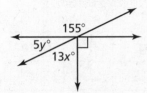

9. The measures of two adjacent angles have a ratio of 3 : 5. The sum of the measures of the two adjacent angles is 120°. What is the measure of the larger angle?

10. Let $x$ be an angle measure. Let $c$ be the measure of the complement of the angle and let $s$ be the measure of the supplement of the angle.

   a. Write an equation involving $c$ and $x$.

   b. Write an equation involving $s$ and $x$.

# Chapter Self-Assessment

Use the scale below to rate your understanding of the learning target and the success criteria.

**1** I do not understand.   **2** I can do it with help.   **3** I can do it on my own.   **4** I can teach someone else.

| | Rating | Date |
|---|---|---|
| **9.1 Circles and Circumference** | | |
| **Learning Target:** Find the circumference of a circle. | 1  2  3  4 | |
| I can explain the relationship between the diameter and circumference of a circle. | 1  2  3  4 | |
| I can use a formula to find the circumference of a circle. | 1  2  3  4 | |
| **9.2 Areas of Circles** | | |
| **Learning Target:** Find the area of a circle. | 1  2  3  4 | |
| I can estimate the area of a circle. | 1  2  3  4 | |
| I can use a formula to find the area of a circle. | 1  2  3  4 | |
| **9.3 Perimeters and Areas of Composite Figures** | | |
| **Learning Target:** Find perimeters and areas of composite figures. | 1  2  3  4 | |
| I can use a grid to estimate perimeters and areas. | 1  2  3  4 | |
| I can identify the shapes that make up a composite figure. | 1  2  3  4 | |
| I can find the perimeters and areas of shapes that make up composite figures. | 1  2  3  4 | |

# Chapter Self-Assessment (continued)

| | Rating | Date |
|---|---|---|
| **9.4 Constructing Polygons** | | |
| **Learning Target:** Construct a polygon with given measures. | 1   2   3   4 | |
| I can use technology to draw polygons. | 1   2   3   4 | |
| I can determine whether given measures result in one triangle, many triangles, or no triangle. | 1   2   3   4 | |
| I can draw polygons given angle measures or side lengths. | 1   2   3   4 | |
| **9.5 Finding Unknown Angle Measures** | | |
| **Learning Target:** Use facts about angle relationships to find unknown angle measures. | 1   2   3   4 | |
| I can identify adjacent, complementary, supplementary, and vertical angles. | 1   2   3   4 | |
| I can use equations to find unknown angle measures. | 1   2   3   4 | |
| I can find unknown angle measures in real-life situations. | 1   2   3   4 | |

## Chapter 10 Review & Refresh

**Find the area of the square or rectangle.**

**1.**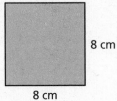
8 cm
8 cm

**2.**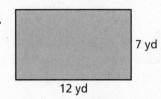
7 yd
12 yd

**3.**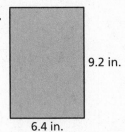
9.2 in.
6.4 in.

**4.**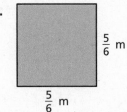
$\frac{5}{6}$ m
$\frac{5}{6}$ m

**5.**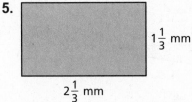
$1\frac{1}{3}$ mm
$2\frac{1}{3}$ mm

**6.**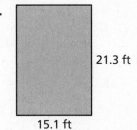
21.3 ft
15.1 ft

**7.** An artist buys a square canvas with a side length of 2.5 feet. What is the area of the canvas?

**Chapter 10** **Review & Refresh** (continued)

**Find the area of the triangle.**

8.

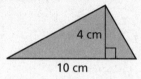

4 cm

10 cm

9.

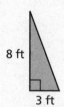

8 ft

3 ft

10.

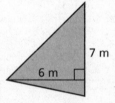

7 m

6 m

11.

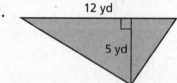

12 yd

5 yd

12.

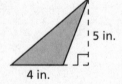

5 in.

4 in.

13.

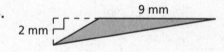

9 mm

2 mm

14. A spirit banner for a pep rally has the shape of a triangle. The base of the banner is 8 feet and the height is 6 feet. Find the area of the banner.

# 10.1 Surface Areas of Prisms
**For use with Exploration 10.1**

**Learning Target:** Find the surface area of a prism.

**Success Criteria:**
- I can use a formula to find the surface area of a prism.
- I can find the lateral surface area of a prism.

---

**1  EXPLORATION: Writing a Formula for Surface Area**

**Work with a partner.**

**a.** Use the diagrams to write a formula for the surface area of a rectangular prism. Explain your reasoning.

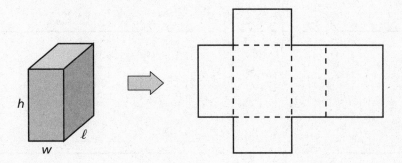

**b.** Choose dimensions for a rectangular prism. Then draw the prism and use your formula in part (a) to find the surface area.

---

**10.1** **Surface Areas of Prisms** (continued)

---

**2** **EXPLORATION:** Surface Area of Prisms

**Work with a partner.**

a. Identify the solid represented by the net (*). Then find the surface area of the solid.

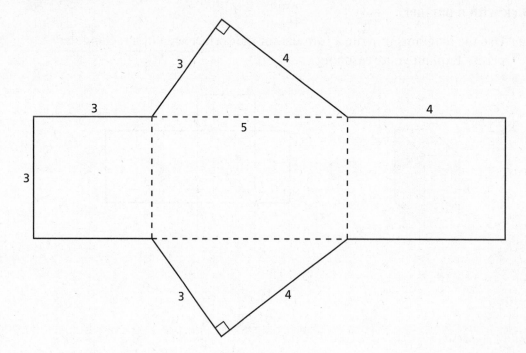

b. Describe a method for finding the surface area of a prism.

* Net is available in the back of the Student Journal.

#  Notetaking with Vocabulary

**Vocabulary:**

**Notes:**

# 10.1 Self-Assessment

Use the scale below to rate your understanding of the learning target and the success criteria.

| **1** | **2** | **3** | **4** |
|---|---|---|---|
| I do not understand. | I can do it with help. | I can do it on my own. | I can teach someone else. |

| | Rating | Date |
|---|---|---|
| **10.1 Surface Areas of Prisms** | | |
| **Learning Target:** Find the surface area of a prism. | 1   2   3   4 | |
| I can use a formula to find the surface area of a prism. | 1   2   3   4 | |
| I can find the lateral surface area of a prism. | 1   2   3   4 | |

**Big Ideas Math: Modeling Real Life Grade 7 Accelerated** **225**
Student Journal

Name _____ Date _____

## 10.1 Practice

**Find the surface area of the prism.**

1.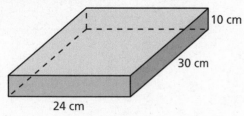

10 cm
30 cm
24 cm

2.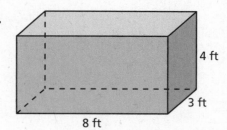

4 ft
3 ft
8 ft

3.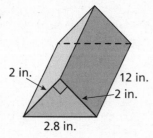

2 in.
12 in.
2 in.
2.8 in.

4.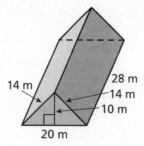

14 m
28 m
14 m
10 m
20 m

5.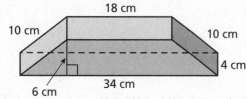

18 cm
10 cm
10 cm
4 cm
34 cm
6 cm

6.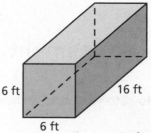

6 ft
16 ft
6 ft

7. A graphing calculator is in the approximate shape of a rectangular prism.

   a. Estimate the total surface area of the calculator.

   b. The window of the calculator is 6.5 centimeters long and 4.5 centimeters wide. Estimate the surface area of the graphing calculator without the window.

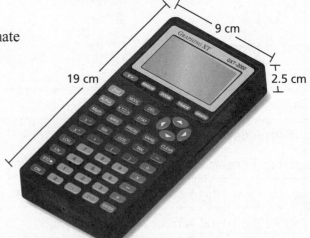

9 cm
19 cm
2.5 cm

8. The least amount of wrapping paper needed to wrap a cube-shaped gift is 150 square inches. How long is one side of the gift?

## 10.2 Surface Areas of Cylinders
**For use with Exploration 10.2**

**Learning Target:** Find the surface area of a cylinder.

**Success Criteria:** • I can use a formula to find the surface area of a cylinder.
• I can find the lateral surface area of a cylinder.

---

**1 EXPLORATION:** Finding the Surface Area of a Cylinder

**Work with a partner.**

   **a.** Make a net for a can. Name each shape in the net.

   **b.** How are the dimensions of the paper related to the dimensions of the can?

---

**Big Ideas Math: Modeling Real Life Grade 7 Accelerated**

**10.2** **Surface Areas of Cylinders** (continued)

   **c.** Write a formula that represents the surface area of a cylinder with a height of $h$ and bases with a radius of $r$.

   **d.** Estimate the dimensions of a can of tuna and a can of soup. Then use your formula in part (c) to estimate the surface area of each can.

## 10.2 Notetaking with Vocabulary

**Vocabulary:**

**Notes:**

## 10.2 Self-Assessment

Use the scale below to rate your understanding of the learning target and the success criteria.

| *1* | *2* | *3* | *4* |
|---|---|---|---|
| I do not understand. | I can do it with help. | I can do it on my own. | I can teach someone else. |

|  | Rating | Date |
|---|---|---|
| **10.2 Surface Areas of Cylinders** | | |
| **Learning Target:** Find the surface area of a cylinder. | 1  2  3  4 | |
| I can use a formula to find the surface area of a cylinder. | 1  2  3  4 | |
| I can find the lateral surface area of a cylinder. | 1  2  3  4 | |

Name_____ Date_____

## 10.2 Practice

**Find the surface area of the cylinder. Round your answer to the nearest tenth if necessary.**

1.

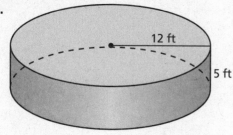

12 ft

5 ft

2.

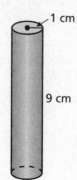

1 cm

9 cm

**Find the lateral surface area of the cylinder. Round your answer to the nearest tenth if necessary.**

3.

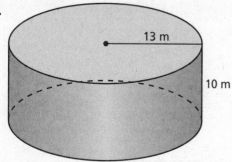

13 m

10 m

4.

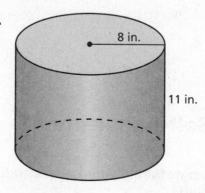

8 in.

11 in.

5. A quarter is worth $0.25 and a half dollar is worth $0.50.

   a. A quarter has a diameter of $\frac{15}{16}$ inch and a height of $\frac{1}{16}$ inch. Find the surface area of a quarter. Round your answer to the nearest hundredth.

   b. A half dollar has a diameter of $\frac{9}{8}$ inches and a height of $\frac{3}{32}$ inch. Find the surface area of a half dollar. Round your answer to the nearest hundredth.

   c. Show that the value of the coin is not proportional to the surface area of the coin.

   d. If the values of the coins were proportional to the surface areas of the coins, what would be the surface area of the half dollar? Round your answer to the nearest hundredth.

# 10.3 Surface Areas of Pyramids
### For use with Exploration 10.3

**Learning Target:** Find the surface area of a pyramid.

**Success Criteria:** • I can use a net to find the surface area of a regular pyramid.
• I can find the lateral surface area of a regular pyramid.

## 1 EXPLORATION: Making a Scale Model

**Work with a partner. Each pyramid listed below has a square base.**

Cheops Pyramid in Egypt

Side ≈ 230 m, Slant height ≈ 186 m

Louvre Pyramid in Paris

Side ≈ 35 m, Slant height ≈ 28 m

    a. Draw a net for a scale model of one of the pyramids. Describe the scale factor.

**10.3** Surface Areas of Pyramids (continued)

**b.** Find the lateral surface area of the real-life pyramid that you chose in part (a). Explain how you found your answer.

**c.** Draw a net for a pyramid with a non-rectangular base and find its lateral surface area. Explain how you found your answer.

# 10.3 Notetaking with Vocabulary

**Vocabulary:**

**Notes:**

# 10.3 Self-Assessment

Use the scale below to rate your understanding of the learning target and the success criteria.

| **1** | **2** | **3** | **4** |
|---|---|---|---|
| I do not understand. | I can do it with help. | I can do it on my own. | I can teach someone else. |

| | Rating | Date |
|---|---|---|
| **10.3 Surface Areas of Pyramids** | | |
| **Learning Target:** Find the surface area of a pyramid. | 1  2  3  4 | |
| I can use a net to find the surface area of a regular pyramid. | 1  2  3  4 | |
| I can find the lateral surface area of a regular pyramid. | 1  2  3  4 | |

## 10.3 Practice

**Find the surface area of the regular pyramid.**

**1.**

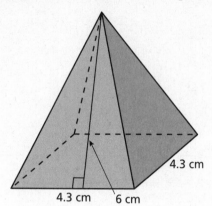

4.3 cm

4.3 cm    6 cm

**2.**

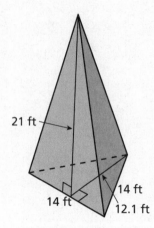

21 ft

14 ft
14 ft    12.1 ft

3. Researchers have determined that a hip roof offers the most protection to a house during a hurricane.

   **a.** The house has a square base with a side length of 50 feet. The house has a variation of a hip roof in the shape of a regular pyramid with a square base. The roof extends 1 foot beyond the walls of the house on all sides. What is the length of each side of the base of the roof?

   **b.** The slant height of the roof is 35 feet. Find the sum of the areas of the lateral faces of the pyramid.

   **c.** A metal roof covering offers the most protection to a house during a hurricane. The cost of installing metal roof covering is $350 for every 100 square feet of roof. What is the cost of installing a metal roof covering on the house?

4. The surface area of a regular triangular pyramid is 197.1 square meters. The slant height is 12 meters. The area of the base is 35.1 square meters. What is the height of the triangular base?

5. The surface area of a regular pentagonal pyramid is 125 square yards. The base length is 5 yards. The area of the base is 37.5 square yards. What is the slant height of the pyramid?

# 10.4 Volumes of Prisms

**For use with Exploration 10.4**

**Learning Target:** Find the volume of a prism.

**Success Criteria:**
- I can use a formula to find the volume of a prism.
- I can use the formula for the volume of a prism to find a missing dimension.

---

### 1 EXPLORATION: Finding a Formula for Volume

**Work with a partner.**

a. In the figures shown, each cube has a volume of 1 cubic unit. Compare the volume $V$ (in cubic units) of each rectangular prism to the area $B$ (in square units) of its base. What do you notice?

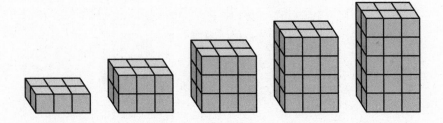

---

**10.4** **Volumes of Prisms** (continued)

**b.** Repeat part (a) using the prisms below.

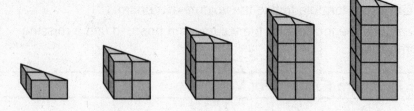

**c.** Use what you learned in parts (a) and (b) to write a formula that gives the volume of any prism.

## 10.4 Notetaking with Vocabulary

**Vocabulary:**

**Notes:**

## 10.4 Self-Assessment

**Use the scale below to rate your understanding of the learning target and the success criteria.**

**1** I do not understand.  **2** I can do it with help.  **3** I can do it on my own.  **4** I can teach someone else.

|  | Rating | Date |
|---|---|---|
| **10.4 Volumes of Prisms** | | |
| **Learning Target:** Find the volume of a prism. | 1   2   3   4 | |
| I can use a formula to find the volume of a prism. | 1   2   3   4 | |
| I can use the formula for the volume of a prism to find a missing dimension. | 1   2   3   4 | |

# 10.4 Practice

**Find the volume of the prism.**

1. 10 cm
   20 cm
   60 cm

2. 10 m
   15 m
   25 m

3. A triangular prism has the measurements as shown.

   a. Find the volume of the prism.

   1.2 yd
   6 yd
   3 yd

   b. Find the volume of the triangular prism if the height of the triangular base is divided by 2.

   c. Explain the relationship between the volume in part (a) and the volume in part (b).

   d. Find the volume of the triangular prism if the height of the prism is divided by 2.

   e. Is the relationship between the volume in part (a) and the volume in part (d) the same as the relationship between the volume in part (a) and the volume in part (b)? Explain.

4. A water tank has the measurements shown. It is 30% full. How many gallons of water are needed to fill the water tank? (100 gal = 0.37854 $m^3$)

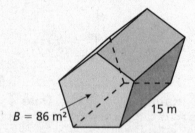

   $B = 86 \ m^2$     15 m

5. A mailbox is in the shape of a prism. The area of the base is 52 square inches and the height is 18 inches. What is the volume of the mailbox?

6. A chicken broth container is in the shape of a rectangular prism, with a length of 9.5 centimeters, a width of 6 centimeters, and a height of 16.5 centimeters. The container is 90% full. How many liters of chicken broth are in the container? (1 L = 1000 $cm^3$). Round your answer to the nearest hundredth.

7. How many cubic feet are in a cubic yard? Use a sketch to explain your reasoning.

# 10.5 Volumes of Pyramids
**For use with Exploration 10.5**

**Learning Target:** Find the volume of a pyramid.

**Success Criteria:** • I can use a formula to find the volume of a pyramid.
• I can use the volume of a pyramid to solve a real-life problem.

---

**1 EXPLORATION:** Finding a Formula for the Volume of a Pyramid

**Work with a partner. Draw the two nets (*) on cardboard and cut them out.
Fold and tape the nets to form an open cube and an open square pyramid.
Both figures should have the same size square base and the same height.**

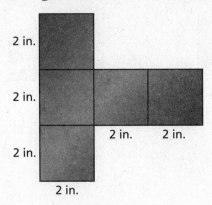

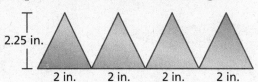

a. Compare the volumes of the figures. What do you notice?

b. Use your observations in part (a) to write a formula for the volume of a pyramid.

* Nets are available in the back of the Student Journal.

---

**Big Ideas Math: Modeling Real Life Grade 7 Accelerated** **239**
Student Journal

**10.5** **Volumes of Pyramids** (continued)

c. The rectangular prism below can be cut to form three pyramids. Use your formula in part (b) to show that the sum of the volumes of the three pyramids is equal to the volume of the prism.

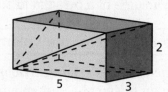

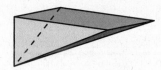  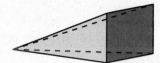  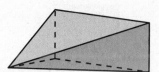

# 10.5 Notetaking with Vocabulary

**Vocabulary:**

**Notes:**

# 10.5 Self-Assessment

Use the scale below to rate your understanding of the learning target and the success criteria.

| **1** | **2** | **3** | **4** |
|---|---|---|---|
| I do not understand. | I can do it with help. | I can do it on my own. | I can teach someone else. |

| | Rating | Date |
|---|---|---|
| **10.5 Volumes of Pyramids** | | |
| **Learning Target:** Find the volume of a pyramid. | 1  2  3  4 | |
| I can use a formula to find the volume of a pyramid. | 1  2  3  4 | |
| I can use the volume of a pyramid to solve a real-life problem. | 1  2  3  4 | |

## 10.5 Practice

**Find the volume of the pyramid.**

**1.**

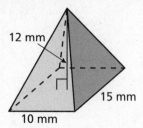

12 mm
15 mm
10 mm

**2.**

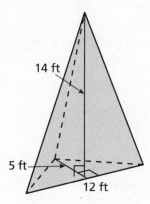

14 ft
5 ft
12 ft

**3.** An amusement park sells popcorn in pyramid-shaped containers. The park charges $0.50 for each 100 cubic centimeters of popcorn. If you purchase the large container, you receive a 25% discount. What is the price for each container? Explain your reasoning.

Pyramid A

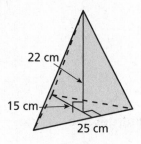

22 cm
15 cm
25 cm

Pyramid B

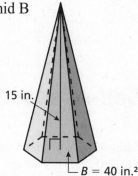

15 in.
$B = 40$ in.²

**4** A pyramid has a rectangular base with length of 15 feet and a width of 8 feet. The height of the pyramid is 10 feet.

**a.** Find the volume of the pyramid.

**b.** When the pyramid was being built, the original base was dropped and it split in two pieces. Each piece was a triangle with a base of 15 feet and a height of 8 feet. Pyramids were made with these two bases, each with a height of 10 feet. Find the combined volume of the two pyramids.

**c.** Is the combined volume *greater than*, *less than*, or the *same as* the volume of the pyramid. Explain your reasoning.

**5.** A triangular pyramid has a volume of 350 cubic centimeters and a height of 10 centimeters. Find one possible set of dimensions of the base.

# 10.6 Cross Sections of Three-Dimensional Figures

**For use with Exploration 10.6**

**Learning Target:** Describe the cross sections of a solid.

**Success Criteria:**
- I can explain the meaning of a cross section.
- I can describe cross sections of prisms and pyramids.
- I can describe cross sections of cylinders and cones.

---

**1 EXPLORATION: Describing Cross Sections**

**Work with a partner. A baker is thinking of different ways to slice zucchini bread that is in the shape of a rectangular prism. The shape that is formed by the cut is called a *cross section*.**

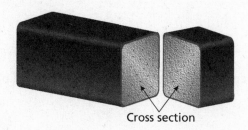

Cross section

a. What is the shape of the cross section when the baker slices the bread vertically, as shown above?

b. What is the shape of the cross section when the baker slices the bread horizontally?

---

**10.6** **Cross Sections of Three-Dimensional Figures** (continued)

**c.** What is the shape of the cross section when the baker slices off a corner of the bread?

**d.** Is it possible to obtain a cross section that is a trapezoid? Explain.

**e.** Name at least 3 cross sections that are possible to obtain from a rectangular pyramid. Explain your reasoning.

# 10.6 Notetaking with Vocabulary

**Vocabulary:**

**Notes:**

# 10.6 Self-Assessment

Use the scale below to rate your understanding of the learning target and the success criteria.

| 1 | 2 | 3 | 4 |
|---|---|---|---|
| I do not understand. | I can do it with help. | I can do it on my own. | I can teach someone else. |

| | Rating | Date |
|---|---|---|
| **10.6 Cross Sections of Three-Dimensional Figures** | | |
| **Learning Target:** Describe the cross sections of a solid. | 1  2  3  4 | |
| I can explain the meaning of a cross section. | 1  2  3  4 | |
| I can describe cross sections of prisms and pyramids. | 1  2  3  4 | |
| I can describe cross sections of cylinders and cones. | 1  2  3  4 | |

Name _____ Date _____

## 10.6 Practice

**Describe the intersection of the plane and the solid.**

1.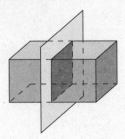

2.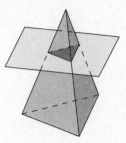

3.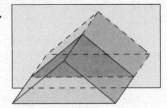

4.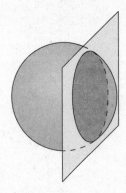

5. Describe the cross section of a plane and a cylinder when the plane is parallel to the base of the cylinder. Draw the solid and the cross section.

6. Describe the cross section of a plane and cone when the plane is perpendicular to the base of the cone and passes through the vertex point of the cone. Draw the solid and the cross section.

**Determine whether the given intersection is possible. If so, draw the solid and the cross section.**

7. The intersection of a plane and a sphere is a point.

8. The intersection of a plane and a cylinder is a triangle.

9. The intersection of a plane and a cylinder is a rectangle.

10. The intersection of a plane and a triangular pyramid is a square.

Name_____ Date_____

# Chapter Self-Assessment

Use the scale below to rate your understanding of the learning target and the success criteria.

| 1 | 2 | 3 | 4 |
|---|---|---|---|
| I do not understand. | I can do it with help. | I can do it on my own. | I can teach someone else. |

| | Rating | Date |
|---|---|---|
| **10.1 Surface Areas of Prisms** | | |
| **Learning Target:** Find the surface area of a prism. | 1  2  3  4 | |
| I can use a formula to find the surface area of a prism. | 1  2  3  4 | |
| I can find the lateral surface area of a prism. | 1  2  3  4 | |
| **10.2 Surface Areas of Cylinders** | | |
| **Learning Target:** Find the surface area of a cylinder. | 1  2  3  4 | |
| I can use a formula to find the surface area of a cylinder. | 1  2  3  4 | |
| I can find the lateral surface area of a cylinder. | | |
| **10.3 Surface Areas of Pyramids** | | |
| **Learning Target:** Find the surface area of a pyramid. | 1  2  3  4 | |
| I can use a net to find the surface area of a regular pyramid. | 1  2  3  4 | |
| I can find the lateral surface area of a regular pyramid. | 1  2  3  4 | |
| **10.4 Volumes of Prisms** | | |
| **Learning Target:** Find the volume of a prism. | 1  2  3  4 | |
| I can use a formula to find the volume of a prism. | 1  2  3  4 | |
| I can use the formula for the volume of a prism to find a missing dimension. | 1  2  3  4 | |

**Chapter 10** Chapter Self-Assessment (continued)

| | Rating | Date |
|---|---|---|
| **10.5 Volumes of Pyramids** | | |
| **Learning Target:** Find the volume of a pyramid. | 1  2  3  4 | |
| I can use a formula to find the volume of a pyramid. | 1  2  3  4 | |
| I can use the volume of a pyramid to solve a real-life problem. | 1  2  3  4 | |
| **10.6 Cross Sections of Three-Dimensional Figures** | | |
| **Learning Target:** Describe the cross sections of a solid. | 1  2  3  4 | |
| I can explain the meaning of a cross section. | 1  2  3  4 | |
| I can describe cross sections of prisms and pyramids. | 1  2  3  4 | |
| I can describe cross sections of cylinders and cones. | 1  2  3  4 | |

Name_____ Date _____

**Reflect the point in (a) the x-axis and (b) the y-axis.**

**1.** $(1, 1)$

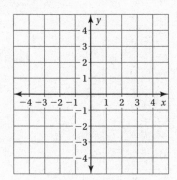

**2.** $(-2, -4)$

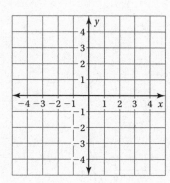

**3.** $(-3, 3)$

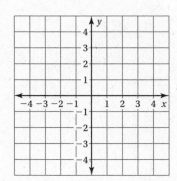

**4.** $(4, -3)$

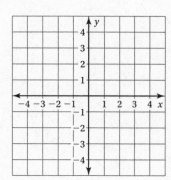

**5.** $(-1, 2)$

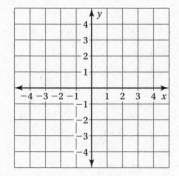

**6.** $(3, 2)$

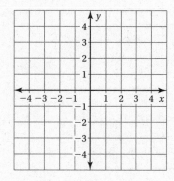

**Chapter 11** **Review & Refresh** (continued)

**Draw the polygon with the given vertices in a coordinate plane.**

**7.** $A(2, 2), B(2, 7), C(6, 7), D(6, 2)$

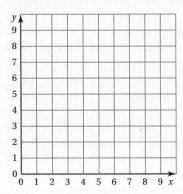

**8.** $E(3, 8), F(3, 1), G(6, 1), H(6, 8)$

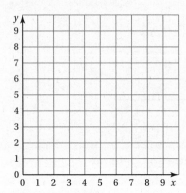

**9.** $I(7, 6), J(5, 2), K(2, 4)$

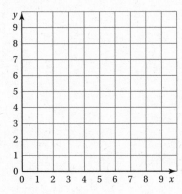

**10.** $L(1, 5), M(1, 2), N(8, 2)$

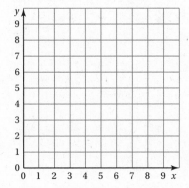

**11.** $O(3, 7), P(6, 7), Q(9, 3), R(1, 3)$

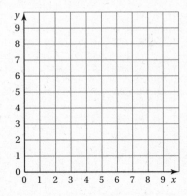

**12.** $S(9, 9), T(7, 1), U(2, 4), V(4, 7)$

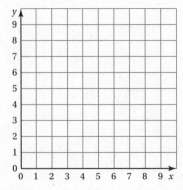

## 11.1 Translations
### For use with Exploration 11.1

**Learning Target:** Translate figures in the coordinate plane.

**Success Criteria:**
- I can identify a translation.
- I can find the coordinates of a translated figure.
- I can use coordinates to translate a figure.

---

**1 EXPLORATION: Sliding Figures**

**Work with a partner.**

a. For each figure below, draw the figure in a coordinate plane. Then copy the figure onto a piece of transparent paper and slide the copy to a new location in the coordinate plane. Describe the location of the copy compared to the location of the original.

- point
- line segment
- line

- triangle
- rectangle

Name _____ Date _____

**11.1** **Translations** (continued)

**b.** When you slide figures, what do you notice about sides, angles, and parallel lines?

**c.** Describe the location of each point below compared to the point $A(x, y)$.

$$B(x + 1, y + 2) \qquad C(x - 3, y + 4)$$

$$D(x - 2, y + 3) \qquad E(x + 4, y - 1)$$

**d.** You copy a point with coordinates $(x, y)$ and slide it horizontally $a$ units and vertically $b$ units. What are the coordinates of the copy?

Name_____ Date_____

**Vocabulary:**

**Notes:**

## 11.1 Self-Assessment

Use the scale below to rate your understanding of the learning target and the success criteria.

| 1 | 2 | 3 | 4 |
|---|---|---|---|
| I do not understand. | I can do it with help. | I can do it on my own. | I can teach someone else. |

|  | Rating | Date |
|---|---|---|
| **11.1 Translations** | | |
| **Learning Target:** Translate figures in the coordinate plane. | 1  2  3  4 | |
| I can identify a translation. | 1  2  3  4 | |
| I can find the coordinates of a translated figure. | 1  2  3  4 | |
| I can use coordinates to translate a figure. | 1  2  3  4 | |

# 11.1 Practice

**Tell whether the right figure is a translation of the left figure.**

1.

2.
 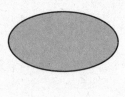

3. Translate the figure 5 units right and 1 unit up. What are the coordinates of the image?

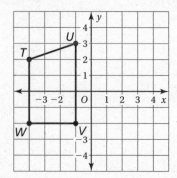

**Describe the translation of the point to its image.**

4. $(1, 5) \rightarrow (-1, 1)$

5. $(-2, -3) \rightarrow (-2, 4)$

6. A square is translated 3 units left and 5 units down. Then the image is translated 4 units right and 2 units down.

   a. Describe the translation of the original square to the ending position.

   b. Describe the translation of the ending position to the original square.

7. You rearrange your bedroom. Tell whether each move is an example of a translation. Explain your reasoning.

   a. You slide your bed 1 foot along the wall.

   b. You move your desk and chair to the opposite wall.

   c. You move your bed stand to the other side of the bed.

## 11.2 Reflections
**For use with Exploration 11.2**

**Learning Target:** Reflect figures in the coordinate plane.

**Success Criteria:**
- I can identify a reflection.
- I can find the coordinates of a figure reflected in an axis.
- I can use coordinates to reflect a figure in the *x*- or *y*-axis.

---

**1  EXPLORATION: Reflecting Figures**

**Work with a partner.**

**a.** For each figure below, draw the figure in the coordinate plane. Then copy the axes and the figure onto a piece of transparent paper. Flip the transparent paper and align the origin and the axes with the coordinate plane. For each pair of figures, describe the line of symmetry.

- point
- line segment
- line

- triangle
- rectangle

**11.2** **Reflections** (continued)

**b.** When you reflect figures, what do you notice about sides, angles, and parallel lines?

**c.** Describe the relationship between each point below and the point $A(4, 7)$ in terms of reflections.

$B(-4, 7)$                    $C(4, -7)$                    $D(-4, -7)$

**d.** A point with coordinates $(x, y)$ is reflected in the $x$-axis. What are the coordinates of the image?

**e.** Repeat part (d) when the point is reflected in the $y$-axis.

Name_____ Date_____

## 11.2 Notetaking with Vocabulary

**Vocabulary:**

**Notes:**

## 11.2 Self-Assessment

Use the scale below to rate your understanding of the learning target and the success criteria.

| *1* | *2* | *3* | *4* |
|---|---|---|---|
| I do not understand. | I can do it with help. | I can do it on my own. | I can teach someone else. |

|  | Rating | Date |
|---|---|---|
| **11.2 Reflections** | | |
| **Learning Target:** Reflect figures in the coordinate plane. | 1   2   3   4 | |
| I can identify a reflection. | 1   2   3   4 | |
| I can find the coordinates of a figure reflected in an axis. | 1   2   3   4 | |
| I can use coordinates to reflect a figure in the *x*- or *y*-axis. | 1   2   3   4 | |

Name _____ Date _____

**Tell whether one figure is a reflection of the other figure.**

1.

2.

**Draw the figure and its reflection in the *x*-axis. Identify the coordinates of the image.**

3. $K(-3,3), L(-2,1), M(1,2), N(2,5)$

4. $O(-2,-1), P(-1,-3), Q(1,-4), R(3,-1)$

**Draw the figure and its reflection in the *y*-axis. Identify the coordinates of the image.**

5. $B(2,-3), C(3,1), D(5,3), E(3,0)$

6. $G(-5,-5), H(-3,-1), I(-2,4), J(-1,-1)$

7. What does the word "pop" spell when it is reflected in a horizontal line?

**The coordinates of a point and its image after a reflection are given. Is the reflection in the *x*-axis or *y*-axis? Explain your reasoning.**

8. $(0,3) \rightarrow (0,-3)$

9. $(1,5) \rightarrow (-1,5)$

10.
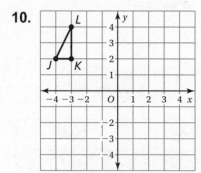

a. The graph shows $\triangle JKL$. Reflect the triangle in the *x*-axis. Then reflect the image in the *y*-axis. Graph the resulting triangle.

b. If the triangle is first reflected in the *y*-axis and then reflected in the *x*-axis, how does this change the resulting triangle? Explain your reasoning.

11. $\triangle ABC$ has vertices $A(-2,-1), B(4,2),$ and $C(2,-2)$.

a. Reflect $\triangle ABC$ in the *x*-axis, giving $\triangle A'B'C'$. Then reflect $\triangle A'B'C'$ in the *y*-axis. What are the coordinates of the resulting triangle?

b. How are the *x*- and *y*-coordinates of the resulting triangle related to the *x*- and *y*-coordinates of $\triangle ABC$?

## 11.3 Rotations
### For use with Exploration 11.3

**Learning Target:** Rotate figures in the coordinate plane.

**Success Criteria:**
- I can identify a rotation.
- I can find the coordinates of a figure rotated about the origin.
- I can use coordinates to rotate a figure about the origin.

---

**1 EXPLORATION: Rotating Figures**

**Work with a partner.**

a. For each figure below, draw the figure in the coordinate plane. Then copy the axes and figure onto a piece of transparent paper. Turn the transparent paper and align the origin and the axes with the coordinate plane. For each pair of figures, describe the angle of rotation.

- point
- line segment
- line

- triangle
- rectangle

## 11.3 Rotations (continued)

**b.** When you rotate figures, what do you notice about sides, angles, and parallel lines?

**c.** Describe the relationship between each point below and the point $A(3, 6)$ in terms of rotations.

$B(-3, -6)$                  $C(6, -3)$                  $D(-6, 3)$

**d.** What are the coordinates of a point $P(x, y)$ after a rotation $90°$ counterclockwise about the origin? $180°$? $270°$?

## 11.3 Notetaking with Vocabulary

**Vocabulary:**

**Notes:**

## 11.3 Self-Assessment

Use the scale below to rate your understanding of the learning target and the success criteria.

| 1 | 2 | 3 | 4 |
|---|---|---|---|
| I do not understand. | I can do it with help. | I can do it on my own. | I can teach someone else. |

|  | Rating | Date |
|---|---|---|
| **11.3 Rotations** | | |
| **Learning Target:** Rotate figures in the coordinate plane. | 1   2   3   4 | |
| I can identify a rotation. | 1   2   3   4 | |
| I can find the coordinates of a figure rotated about the origin. | 1   2   3   4 | |
| I can use coordinates to rotate a figure about the origin. | 1   2   3   4 | |

# 11.3 Practice

Tell whether the dashed figure is a rotation of the solid figure about the origin. If so, give the angle and direction of rotation.

1.
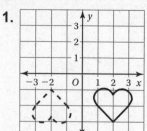

2.
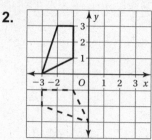

The vertices of a trapezoid are $A(1,1)$, $B(2,2)$, $C(4,2)$, and $D(5,1)$. Rotate the trapezoid as described. Find the coordinates of the image.

3. 90° clockwise about the origin

4. 270° counterclockwise about the origin

5. 90° clockwise about vertex $A$

6. 180° about vertex $D$

Determine whether the figure has rotational symmetry. Explain your reasoning.

7.
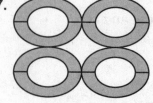

8.

9. You rotate a triangle 270° counterclockwise about the origin. Then you translate its image 2 units right and 1 unit down. The vertices of the final image are $(0,2)$, $(8,-1)$, and $(5,-2)$. What are the vertices of the original triangle?

# 11.4 Congruent Figures
### For use with Exploration 11.4

**Learning Target:** Understand the concept of congruent figures.

**Success Criteria:**
- I can identify congruent figures.
- I can describe a sequence of rigid motions between two congruent figures.

## 1 EXPLORATION: Transforming Figures

**Work with a partner.**

**a.** For each pair of figures whose vertices are given below, draw the figures in a coordinate plane. Then copy one of the figures onto a piece of transparent paper. Use transformations to try to obtain one of the figures from the other figure.

- $A(-5, 1), B(-5, -4), C(-2, -4)$ and $D(1, 4), E(1, -1), F(-2, -1)$

- $G(1, 2), H(2, -6), J(5, 0)$ and $L(-1, -2), M(-2, 6), N(-5, 0)$

- $P(0, 0), Q(2, 2), R(4, -2)$ and $X(0, 0), Y(3, 3), Z(6, -3)$

- $A(0, 4), B(3, 8), C(6, 4), D(3, 0)$ and $F(-4, -3), G(-8, 0), H(-4, 3), J(0, 0)$

- $P(-2, 1), Q(-1, -2), R(1, -2), S(1, 1)$ and $W(7, 1), X(5, -2), Y(3, -2), Z(3, 1)$

**Big Ideas Math: Modeling Real Life Grade 7 Accelerated** **263**

**11.4** **Congruent Figures** (continued)

   **b.** Which pairs of figures in part (a) are identical? Explain your reasoning.

   **c.** Figure A and Figure B are identical. Do you think there must be a sequence of transformations that obtains Figure A from Figure B? Explain your reasoning.

**Notetaking with Vocabulary**

**Vocabulary:**

**Notes:**

 **Self-Assessment**

Use the scale below to rate your understanding of the learning target and the success criteria.

| | 1 | 2 | 3 | 4 |
|---|---|---|---|---|
| | I do not understand. | I can do it with help. | I can do it on my own. | I can teach someone else. |

| | Rating | Date |
|---|---|---|
| **11.4 Congruent Figures** | | |
| **Learning Target:** Understand the concept of congruent figures. | 1   2   3   4 | |
| I can identify congruent figures. | 1   2   3   4 | |
| I can describe a sequence of rigid motions between two congruent figures. | 1   2   3   4 | |

## 11.4 Practice

**The figures are congruent. Name the corresponding angles and the corresponding sides.**

1.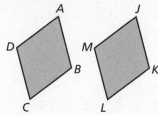

2.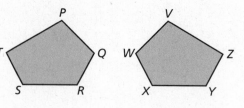

3. The figures are congruent.

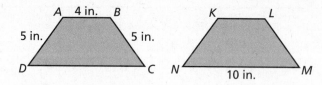

    **a.** What is the length of side *CD*?

    **b.** Which angle of *KLMN* corresponds to ∠*B*?

    **c.** What is the perimeter of *KLMN*?

4. The pentagons are congruent. Determine whether the statement is *true* or *false*. Explain your reasoning.

    **a.** ∠*B* is congruent to ∠*C*.

    **b.** Side *MN* is congruent to side *AE*.

    **c.** ∠*B* corresponds to ∠*O*.

    **d.** Side *BC* is congruent to side *PO*.

    **e.** The sum of the angle measures of *LMNOP* is 540°.

    **f.** The measure of ∠*B* is 120°.

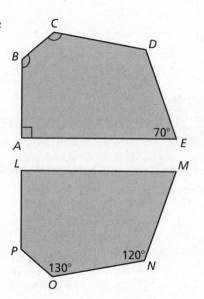

5. △*JKL* is congruent to △*PQR*. Describe a sequence of rigid motions between the figures.

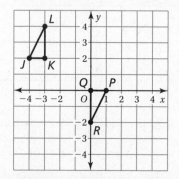

## 11.5 Dilations
### For use with Exploration 11.5

**Learning Target:** Dilate figures in the coordinate plane.

**Success Criteria:**
- I can identify a dilation.
- I can find the coordinates of a figure dilated with respect to the origin.
- I can use coordinates to dilate a figure with respect to the origin.

---

**1 EXPLORATION: Dilating a Polygon**

**Work with a partner. Use geometry software.**

**a.** Draw a polygon in the coordinate plane. Then *dilate* the polygon with respect to the origin. Describe the scale factor of the image.

**b.** Compare the image and the original polygon in part (a). What do you notice about the sides? the angles?

**11.5** **Dilations** (continued)

**c.** Describe the relationship between each point below and the point $A(x, y)$ in terms of dilations.

$B(3x, 3y)$ $\qquad$ $C(5x, 5y)$ $\qquad$ $D(0.5x, 0.5y)$

**d.** What are the coordinates of a point $P(x, y)$ after a dilation with respect to the origin by a scale factor of $k$?

## 11.5 Notetaking with Vocabulary

**Vocabulary:**

**Notes:**

## 11.5 Self-Assessment

Use the scale below to rate your understanding of the learning target and the success criteria.

| 1 | 2 | 3 | 4 |
|---|---|---|---|
| I do not understand. | I can do it with help. | I can do it on my own. | I can teach someone else. |

| | Rating | Date |
|---|---|---|
| **11.5 Dilations** | | |
| **Learning Target:** Dilate figures in the coordinate plane. | 1   2   3   4 | |
| I can identify a dilation. | 1   2   3   4 | |
| I can find the coordinates of a figure dilated with respect to the origin. | 1   2   3   4 | |
| I can use coordinates to dilate a figure with respect to the origin. | 1   2   3   4 | |

# 11.5 Practice

**Tell whether the dashed figure is a dilation of the solid figure.**

1.

2.

**The vertices of a figure are given. Draw the figure and its image after a dilation with the given scale factor. Identify the type of dilation.**

3. $A(3, -1), B(-4, 4), C(-2, -3); k = 5$

4. $D(10, 20), E(-35, 10), F(25, -30), G(5, -20); k = \frac{1}{5}$

**The dashed figure is a dilation of the solid figure. Identify the type of dilation and find the scale factor. Explain your reasoning.**

5.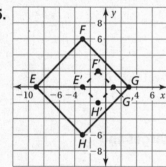

6.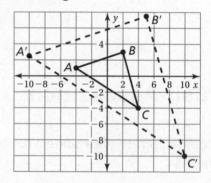

7. A scale factor of 2 is used to find the dilation of a quadrilateral.

   a. What is the sum of the angles in the original quadrilateral?

   b. What is the sum of the angles of the image after the dilation?

   c. What is the difference between the perimeter of the original figure and the perimeter of the image?

8. A triangle is dilated using a scale factor of $\frac{1}{2}$. The image is then dilated using a scale factor of $\frac{1}{3}$. What scale factor could you use to dilate the original triangle to get the final image?

9. A triangle has an area of 3. The triangle is dilated using a scale factor of 3. What is the area of the image after the dilation? Explain your reasoning.

## 11.6 Similar Figures
**For use with Exploration 11.6**

**Learning Target:** Understand the concept of similar figures.

**Success Criteria:**
- I can identify similar figures.
- I can describe a similarity transformation between two similar figures.

**1 EXPLORATION: Transforming Figures**

**Work with a partner. Use geometry software.**

a. For each pair of figures whose vertices are given below, draw the figures in a coordinate plane. Use dilations and rigid motions to try to obtain one of the figures from the other figure.

- $A(-3, 6), B(0, -3), C(3, 6)$ and $G(-1, 2), H(0, -1), J(1, 2)$

- $D(0, 0), E(3, 0), F(3, 3)$ and $L(0, 0), M(0, 6), N(-6, 6)$

- $P(1, 0), Q(4, 2), R(7, 0)$ and $X(-1, 0), Y(-4, 6), Z(-7, 0)$

- $A(-3, 2), B(-1, 2), C(-1, -1), D(-3, -1)$ and
  $F(6, 4), G(2, 4), H(2, -2), J(6, -2)$

- $P(-2, 2), Q(-1, -1), R(1, -1), S(2, 2)$ and
  $W(2, 8), X(3, 3), Y(7, 3), Z(8, 8)$

**11.6** **Similar Figures** (continued)

**b.** Is a scale drawing represented by any of the pairs of figures in part (a)?
Explain your reasoning.

**c.** Figure A is a scale drawing of Figure B. Do you think there must be a
sequence of transformations that obtains Figure A from Figure B? Explain
your reasoning.

## 11.6 Notetaking with Vocabulary

**Vocabulary:**

**Notes:**

## 11.6 Self-Assessment

Use the scale below to rate your understanding of the learning target and the success criteria.

| 1 | 2 | 3 | 4 |
|---|---|---|---|
| I do not understand. | I can do it with help. | I can do it on my own. | I can teach someone else. |

| | Rating | Date |
|---|---|---|
| **11.6 Similar Figures** | | |
| **Learning Target:** Understand the concept of similar figures. | 1   2   3   4 | |
| I can identify similar figures. | 1   2   3   4 | |
| I can describe a similarity transformation between two similar figures. | 1   2   3   4 | |

# 11.6 Practice

1. Draw the figures with the given vertices in a coordinate plane. Which figures are similar? Explain your reasoning.

   Rectangle A: $(0, 0), (3, 0), (3, 2), (0, 2)$

   Rectangle B: $(0, 0), (1, 0), (1, 3), (0, 3)$

   Rectangle C: $(0, 0), (2, 0), (2, -3), (0, -3)$

2. A rectangular index card is 6 inches long and 4 inches wide. A rectangular note card is 1.5 inches long and 1 inch wide. Are the cards similar?

**The two parallelograms are similar. Find the degree measure of the angle. Explain your reasoning.**

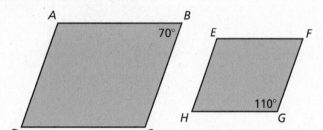

3. $\angle A$

4. $\angle H$

5. $\angle D$

6. $\angle F$

7. Is it possible for the following figures to be similar? Explain.

   a. A stop sign and a speed limit sign

   b. A cell phone and a test paper

   c. A yield sign and a home plate

   d. A laptop and a swimming pool

8. Can you draw two triangles each having two 45° angles and one 90° angle that are *not* similar? Justify your answer.

9. You have a triangle that has side lengths of 6, 9, and 12.

   a. Give the side lengths of a similar triangle that is smaller than the given triangle. Justify your answer.

   b. Give the side lengths of a similar triangle that is larger than the given triangle. Justify your answer.

   c. Each side length is increased by 30%. What are the side lengths of the new triangle? Is the new triangle similar to the original triangle?

# 11.7 Perimeters and Areas of Similar Figures
**For use with Exploration 11.7**

**Learning Target:** Find perimeters and areas of similar figures.

**Success Criteria:**
- I can use corresponding side lengths to compare perimeters of similar figures.
- I can use corresponding side lengths to compare areas of similar figures.
- I can use similar figures to solve real-life problems involving perimeter and area.

---

**1** **EXPLORATION:** Comparing Similar Figures

**Work with a partner. Draw a rectangle in the coordinate plane.**

**a.** Dilate your rectangle using each indicated scale factor $k$. Then complete the table for the perimeter $P$ of each rectangle. Describe the pattern.

| Original Side Lengths | $k = 2$ | $k = 3$ | $k = 4$ | $k = 5$ | $k = 6$ |
|---|---|---|---|---|---|
| $P = $ ___ | | | | | |

**b.** Compare the ratios of the perimeters to the ratios of the corresponding side lengths. What do you notice?

**11.7** **Perimeters and Areas of Similar Figures** (continued)

**c.** Repeat part (a) to complete the table for the area *A* of each rectangle. Describe the pattern.

| Original Side Lengths | $k = 2$ | $k = 3$ | $k = 4$ | $k = 5$ | $k = 6$ |
|---|---|---|---|---|---|
| $A = $ ___ | | | | | |

**d.** Compare the ratios of the areas to the ratios of the corresponding side lengths. What do you notice?

**e.** The rectangles shown are similar. You know the perimeter and the area of the large rectangle and a pair of corresponding side lengths. How can you find the perimeter of the small rectangle? the area of the small rectangle?

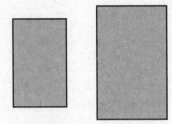

## 11.7 Notetaking with Vocabulary

**Vocabulary:**

**Notes:**

## 11.7 Self-Assessment

**Use the scale below to rate your understanding of the learning target and the success criteria.**

| 1 | 2 | 3 | 4 |
|---|---|---|---|
| I do not understand. | I can do it with help. | I can do it on my own. | I can teach someone else. |

| | Rating | Date |
|---|---|---|
| **11.7 Perimeters and Areas of Similar Figures** | | |
| **Learning Target:** Find perimeters and areas of similar figures. | 1  2  3  4 | |
| I can use corresponding side lengths to compare perimeters of similar figures. | 1  2  3  4 | |
| I can use corresponding side lengths to compare areas of similar figures. | 1  2  3  4 | |
| I can use similar figures to solve real-life problems involving perimeter and area. | 1  2  3  4 | |

## 11.7 Practice

1. The two figures are similar. Find the ratio (small to large) of the perimeters and of the areas.

 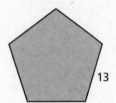

2. The figures are similar. The ratio of the perimeters is 12 : 7. Find $x$.

3. The ratio of the corresponding side lengths of two similar parallelogram signs is 9 : 14.

   a. What is the ratio of the perimeters? Explain.

   b. What is the ratio of the areas? Explain.

   c. One side length of the smaller sign is 45 feet. What is the side length of the corresponding side of the larger sign?

4. A window is put in a door. The window and the door are similar rectangles. The door has a width of 4 feet. The window has a width of 30 inches.

   a. How many times greater is the area of the door than the area of the window?

   b. The area of the door is 32 square feet. What is the area of the window?

   c. What is the perimeter of the window?

5. The area of Circle P is $4\pi$. The area of Circle Q is $25\pi$.

   a. What is the ratio of their areas?

   b. What is the ratio of their radii? Justify your reasoning.

   c. The radius of Circle Q is decreased by 50%. What is the new circumference of Circle Q?

Name_____ Date_____

## Chapter Self-Assessment

Use the scale below to rate your understanding of the learning target and the success criteria.

| **1** | **2** | **3** | **4** |
|---|---|---|---|
| I do not understand. | I can do it with help. | I can do it on my own. | I can teach someone else. |

| | Rating | Date |
|---|---|---|
| **11.1 Translations** | | |
| **Learning Target:** Translate figures in the coordinate plane. | 1  2  3  4 | |
| I can identify a translation. | 1  2  3  4 | |
| I can find the coordinates of a translated figure. | 1  2  3  4 | |
| I can use coordinates to translate a figure. | 1  2  3  4 | |
| **11.2 Reflections** | | |
| **Learning Target:** Reflect figures in the coordinate plane. | 1  2  3  4 | |
| I can identify a reflection. | 1  2  3  4 | |
| I can find the coordinates of a figure reflected in an axis. | 1  2  3  4 | |
| I can use coordinates to reflect a figure in the x- or y-axis. | 1  2  3  4 | |
| **11.3 Rotations** | | |
| **Learning Target:** Rotate figures in the coordinate plane. | 1  2  3  4 | |
| I can identify a rotation. | 1  2  3  4 | |
| I can find the coordinates of a figure rotated about the origin. | 1  2  3  4 | |
| I can use coordinates to rotate a figure about the origin. | 1  2  3  4 | |

# Chapter 11 Chapter Self-Assessment (continued)

| | Rating | Date |
|---|---|---|
| **11.4 Congruent Figures** | | |
| **Learning Target:** Understand the concept of congruent figures. | 1   2   3   4 | |
| I can identify congruent figures. | 1   2   3   4 | |
| I can describe a sequence of rigid motions between two congruent figures. | 1   2   3   4 | |
| **11.5 Dilations** | | |
| **Learning Target:** Dilate figures in the coordinate plane. | 1   2   3   4 | |
| I can identify a dilation. | 1   2   3   4 | |
| I can find the coordinates of a figure dilated with respect to the origin. | 1   2   3   4 | |
| I can use coordinates to dilate a figure with respect to the origin. | 1   2   3   4 | |
| **11.6 Similar Figures** | | |
| **Learning Target:** Understand the concept of similar figures. | 1   2   3   4 | |
| I can identify similar figures. | 1   2   3   4 | |
| I can describe a similarity transformation between two similar figures. | 1   2   3   4 | |
| **11.7 Perimeters and Areas of Similar Figures** | | |
| **Learning Target:** Find perimeters and areas of similar figures. | 1   2   3   4 | |
| I can use corresponding side lengths to compare perimeters of similar figures. | 1   2   3   4 | |
| I can use corresponding side lengths to compare areas of similar figures. | 1   2   3   4 | |
| I can use similar figures to solve real-life problems involving perimeter and area. | 1   2   3   4 | |

**Chapter 12** **Review & Refresh**

Tell whether the angles are *adjacent* or *vertical*. Then find the value of *x*.

1.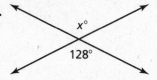

$x°$

$128°$

2.

$x°$

$35°$

3.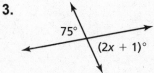

$75°$

$(2x + 1)°$

4.

$4x°$

$2x°$

5. The tree is tilted 14°. Find the value of *x*.

$14°$

$x°$

**Big Ideas Math: Modeling Real Life Grade 7 Accelerated**
Student Journal
**281**

**Chapter 12** **Review & Refresh** (continued)

**Tell whether the angles are *complementary* or *supplementary*. Then find the value of *x*.**

6.

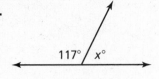

117° $x°$

7.

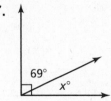

69° $x°$

8.

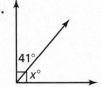

41° $x°$

9.

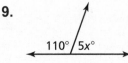

110° $5x°$

10. A tributary joins a river at an angle. Find the value of *x*.

$x°$   127°

## 12.1 Parallel Lines and Transversals
For use with Exploration 12.1

**Learning Target:** Find missing angle measures created by the intersections of lines.

**Success Criteria:** • I can identify congruent angles when a transversal intersects parallel lines.
• I can find angle measures when a transversal intersects parallel lines.

**1 EXPLORATION: Exploring Intersections of Lines**

**Work with a partner. Use geometry software and the lines *A* and *B* shown.**

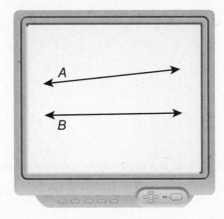

**a.** Are line *A* and line *B* parallel? Explain your reasoning.

**b.** Draw a line *C* that intersects both line *A* and line *B*. What do you notice about the measures of the angles that are created?

**Big Ideas Math: Modeling Real Life Grade 7 Accelerated** **283**
Student Journal

## 12.1  Parallel Lines and Transversals (continued)

c.  Rotate line *A* or line *B* until the angles created by the intersection of line *A* and line *C* are congruent to the angles created by the intersection of line *B* and line *C*. What do you notice about line *A* and line *B*?

d.  Rotate line *C* to create different angle measures. Are the angles that were congruent in part (c) still congruent?

e.  Make a conjecture about the measures of the angles created when a line intersects two parallel lines.

**Notetaking with Vocabulary**

**Vocabulary:**

**Notes:**

**12.1** **Self-Assessment**

Use the scale below to rate your understanding of the learning target and the success criteria.

| *1* | *2* | *3* | *4* |
|---|---|---|---|
| I do not understand. | I can do it with help. | I can do it on my own. | I can teach someone else. |

| | Rating | Date |
|---|---|---|
| **12.1 Parallel Lines and Transversals** | | |
| **Learning Target**: Find missing angle measures created by the intersections of lines. | 1　2　3　4 | |
| I can identify congruent angles when a transversal intersects parallel lines. | 1　2　3　4 | |
| I can find angle measures when a transversal intersects parallel lines. | 1　2　3　4 | |

# 12.1 Practice

**Use the figure to find the measures of the numbered angles. Explain your reasoning.**

**1.**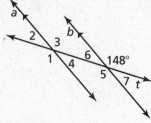

**2.**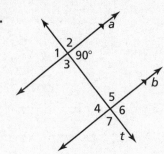

**Complete the statement. Explain your reasoning.**

**3.** If the measure of ∠1 = 130°, then the measure of ∠8 = ____.

**4.** If the measure of ∠5 = 53°, then the measure of ∠3 = ____.

**5.** If the measure of ∠7 = 71°, then the measure of ∠3 = ____.

**6.** If the measure of ∠4 = 65°, then the measure of ∠6 = ____.

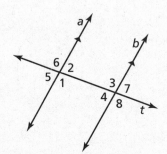

**Using the diagram for angle placement only (the measurement of the angles may change), indicate if the following statements are *always*, *sometimes*, or *never* true. Explain.**

**7.** ∠1 is congruent to ∠3.

**8.** ∠6 is supplementary to ∠8.

**9.** ∠2 is complementary to ∠1.

**10.** ∠8 and ∠5 are vertical angles.

**11.** ∠2 is congruent to ∠8.

**12.** If a transversal intersects two parallel lines, is it possible for all of the angles formed to be acute angles? Explain.

# 12.2 Angles of Triangles
**For use with Exploration 12.2**

**Learning Target:** Understand properties of interior and exterior angles of triangles.

**Success Criteria:**
- I can use equations to find missing angle measures of triangles.
- I can use interior and exterior angles of a triangle to solve real-life problems.

---

**1 EXPLORATION: Exploring Interior and Exterior Angles of Triangles**

**Work with a partner.**

**a.** Draw several triangles using geometry software. What can you conclude about the sums of the angle measures?

**b.** You can extend one side of a triangle to form an *exterior angle,* as shown.

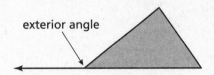

exterior angle

Use geometry software to draw a triangle and an exterior angle. Compare the measure of the exterior angle with the measures of the interior angles. Repeat this process for several different triangles. What can you conclude?

## 12.2 Angles of Triangles (continued)

### 2 EXPLORATION: Using Parallel Lines and Transversals

**Work with a partner. Describe what is shown in the figure below. Then use what you know about parallel lines and transversals to justify your conclusions in Exploration 1.**

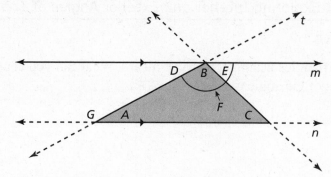

Name_____ Date_____

 **Notetaking with Vocabulary**

**Vocabulary:**

**Notes:**

 **Self-Assessment**

Use the scale below to rate your understanding of the learning target and the success criteria.

| **1** | **2** | **3** | **4** |
|---|---|---|---|
| I do not understand. | I can do it with help. | I can do it on my own. | I can teach someone else. |

|  | Rating | Date |
|---|---|---|
| **12.2 Angles of Triangles** | | |
| **Learning Target:** Understand properties of interior and exterior angles of triangles. | 1   2   3   4 | |
| I can use equations to find missing angle measures of triangles. | 1   2   3   4 | |
| I can use interior and exterior angles of a triangle to solve real-life problems. | 1   2   3   4 | |

Name_____ Date _____

## 12.2 Practice

**Find the measure of the interior angles of the triangle.**

**1.**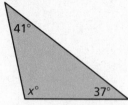

**2.**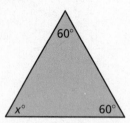

**Find the measure of the exterior angle.**

**3.**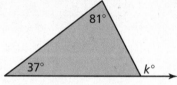

**4.**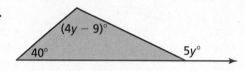

**5.** The ratio of the interior angle measures of a triangle is 1 : 4 : 5. What are the angle measures?

**6.** A right triangle has an exterior angle with a measure of 160°. Can you determine the measures of the interior angles? Explain.

**7.** You are in a sailboat race. The course is triangular, racing A – B – C – Finish. By what angle do the sailboats need to change so that they can make it to the finish line?

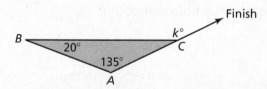

**Determine whether the statement is *always*, *sometimes*, or *never* true. Explain your reasoning.**

**8.** The exterior angles of an equilateral triangle all have the same measure.

**9.** All equilateral triangles have exterior angles with the same measures.

**10.** A triangle has two vertices with an obtuse exterior angle.

# 12.3 Angles of Polygons
### For use with Exploration 12.3

**Learning Target:** Find interior angle measures of polygons.

**Success Criteria:**
- I can explain how to find the sum of the interior angle measures of a polygon.
- I can use an equation to find an interior angle measure of a polygon.
- I can find the interior angle measures of a regular polygon.

---

**1 EXPLORATION:** Exploring Interior Angles of Polygons

**Work with a partner. In parts (a)-(f), use what you know about the interior angle measures of triangles to find the sum of the interior angle measures of each figure.**

a.

b.

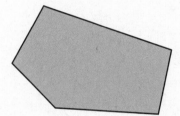

c.

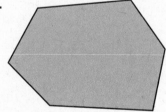

d.

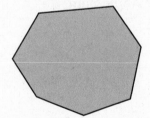

---

**12.3** **Angles of Polygons** (continued)

e.
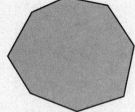

f.
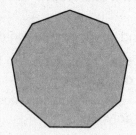

g. Use your results in parts (a)–(f) to complete the table. Then write an equation that represents the sum $S$ of the interior angle measures of a polygon with $n$ sides.

| Number of Sides, $n$ | 3 | 4 | 5 | 6 | 7 | 8 | 9 |
|---|---|---|---|---|---|---|---|
| Number of Triangles | | | | | | | |
| Interior Angle Sum, $S$ | | | | | | | |

## 12.3 Notetaking with Vocabulary

**Vocabulary:**

**Notes:**

## 12.3 Self-Assessment

**Use the scale below to rate your understanding of the learning target and the success criteria.**

| **1** | **2** | **3** | **4** |
|---|---|---|---|
| I do not understand. | I can do it with help. | I can do it on my own. | I can teach someone else. |

|  | Rating | Date |
|---|---|---|
| **12.3 Angles of Polygons** | | |
| **Learning Target:** Find interior angle measures of polygons. | 1   2   3   4 | |
| I can explain how to find the sum of the interior angle measures of a polygon. | 1   2   3   4 | |
| I can use an equation to find an interior angle measure of a polygon. | 1   2   3   4 | |
| I can find the interior angle measures of a regular polygon. | 1   2   3   4 | |

Name _____  Date _____

## 12.3  Practice

**Use triangles to find the sum of the interior angle measures of the polygon.**

1.

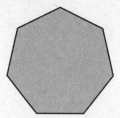

2.

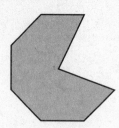

**Find the sum of the interior angle measures of the polygon.**

3.

4.

**Find the value of x.**

5.

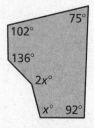

6.

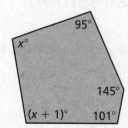

7. The interior angles of a regular polygon each measure 135°. How many sides does the polygon have?

8. Home plate at a baseball field is in the shape of a pentagon. Three of the interior angles are right angles, the other two angles have equal measures. What are the measures of the interior angles of home plate?

9. A Ferris wheel at a carnival is in the shape of a regular dodecagon, a twelve-sided figure.

   a. What is the sum of the interior angles of the Ferris wheel?

   b. What is the measure of each interior angle of the Ferris wheel?

## 12.4 Using Similar Triangles
**For use with Exploration 12.4**

**Learning Target:** Use similar triangles to find missing measures.

**Success Criteria:**
- I can use angle measures to determine whether triangles are similar.
- I can use similar triangles to solve real-life problems.

---

**1 EXPLORATION: Drawing Triangles Given Two Angle Measures**

**Work with a partner. Use geometry software.**

**a.** Draw a triangle that has a 50° angle and a 30° angle. Then draw a triangle that is either larger or smaller that has the same two angle measures. Are the triangles congruent? similar? Explain your reasoning.

**b.** Choose any two angle measures whose sum is less than 180°. Repeat part (a) using the angle measures you chose.

**c.** Compare your results in parts (a) and (b) with other pairs of students. Make a conjecture about two triangles that have two pairs of congruent angles.

## 12.4 Using Similar Triangles (continued)

**2** **EXPLORATION:** Using Indirect Measurement

Work with a partner. Use the fact that two rays from the Sun are parallel to make a plan for how to find the height of the flagpole. Explain your reasoning.

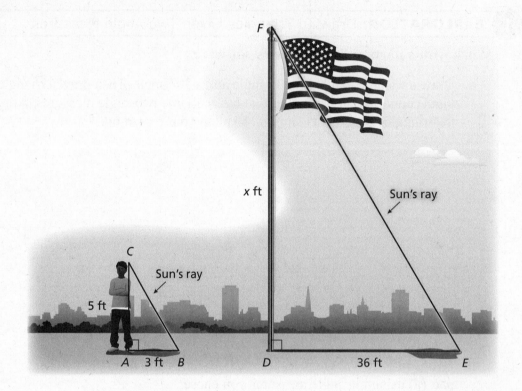

**Notetaking with Vocabulary**

**Vocabulary:**

**Notes:**

**12.4  Self-Assessment**

Use the scale below to rate your understanding of the learning target and the success criteria.

| *1* | *2* | *3* | *4* |
|---|---|---|---|
| I do not understand. | I can do it with help. | I can do it on my own. | I can teach someone else. |

| | Rating | Date |
|---|---|---|
| **12.4 Using Similar Triangles** | | |
| **Learning Target:** Use similar triangles to find missing measures. | 1  2  3  4 | |
| I can use angle measures to determine whether triangles are similar. | 1  2  3  4 | |
| I can use similar triangles to solve real-life problems. | 1  2  3  4 | |

## 12.4 Practice

**Tell whether the triangles are similar. Explain.**

1.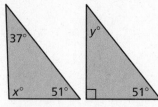

2.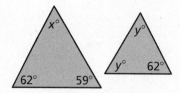

3. The triangles are similar. Find the value of $x$.

4. You can use indirect measurement to estimate the height of a flag pole. First measure your distance from the base of the flag pole and the distance from the ground to a point on the flag pole that you are looking at. Maintaining the same angle of sight, move back until the top of the flag pole is in your line of sight.

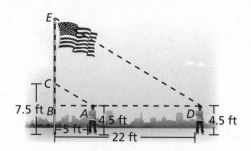

   a. Explain why $\triangle ABC$ and $\triangle DBE$ are similar.

   b. What is the height of the flag pole?

5. You are on a boat in the ocean, at Point $A$. You locate a lighthouse at Point $D$, beyond the line of sight of the marker at point $C$. You drive 0.2 mile west to Point $B$ and then 0.1 mile south to Point $C$. You drive 0.3 mile more to arrive at Point $E$, which is due east of the lighthouse.

   a. Explain why $\triangle ABC$ and $\triangle DEC$ are similar.

   b. What is the distance from Point $E$ to the lighthouse?

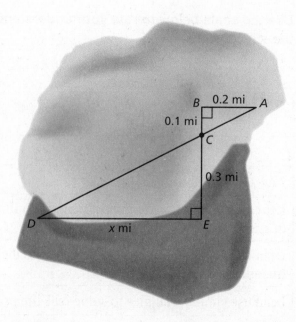

Name_____ Date_____

# Chapter Self-Assessment

**Use the scale below to rate your understanding of the learning target and the success criteria.**

| 1 | 2 | 3 | 4 |
|---|---|---|---|
| I do not understand. | I can do it with help. | I can do it on my own. | I can teach someone else. |

| | Rating | Date |
|---|---|---|
| **12.1 Parallel Lines and Transversals** | | |
| **Learning Target:** Find missing angle measures created by the intersections of lines. | 1  2  3  4 | |
| I can identify congruent angles when a transversal intersects parallel lines. | 1  2  3  4 | |
| I can find angle measures when a transversal intersects parallel lines. | 1  2  3  4 | |
| **12.2 Angles of Triangles** | | |
| **Learning Target:** Understand properties of interior and exterior angles of triangles. | 1  2  3  4 | |
| I can use equations to find missing angle measures of triangles. | 1  2  3  4 | |
| I can use interior and exterior angles of a triangle to solve real-life problems. | 1  2  3  4 | |
| **12.3 Angles of Polygons** | | |
| **Learning Target:** Find interior angle measures of polygons. | 1  2  3  4 | |
| I can explain how to find the sum of the interior angle measures of a polygon. | 1  2  3  4 | |
| I can use an equation to find an interior angle measure of a polygon. | 1  2  3  4 | |
| I can find the interior angle measures of a regular polygon. | 1  2  3  4 | |

## Chapter 12 Chapter Self-Assessment (continued)

|  | Rating | Date |
|---|---|---|
| **12.4 Using Similar Triangles** | | |
| **Learning Target:** Use similar triangles to find missing measures. | 1　2　3　4 | |
| I can use angle measures to determine whether triangles are similar. | 1　2　3　4 | |
| I can use similar triangles to solve real-life problems. | 1　2　3　4 | |

## Chapter 13 Review & Refresh

**Evaluate the expression when $x = \frac{1}{2}$ and $y = -5$.**

**1.** $-2xy$ **2.** $4x^2 - 3y$

**3.** $\dfrac{10y}{12x+4}$ **4.** $11x - 8(x - y)$

**Evaluate the expression when $a = -9$ and $b = -4$.**

**5.** $3ab$ **6.** $a^2 - 2(b + 12)$

**7.** $\dfrac{4b^2}{3b-7}$ **8.** $7b^2 + 5(ab - 6)$

**9.** You go to the movies with five friends. You and one of your friends each buy a ticket and a bag of popcorn. The rest of your friends buy just one ticket each. The expression $4x + 2(x + y)$ represents the situation. Evaluate the expression when tickets cost \$7.25 and a bag of popcorn costs \$3.25.

**Chapter 13** **Review & Refresh** (continued)

**Use the graph to answer the question.**

10. Write the ordered pair that corresponds to Point $D$.

11. Write the ordered pair that corresponds to Point $H$.

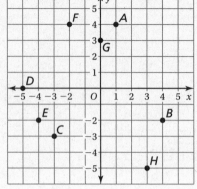

12. Which point is located at $(-2, 4)$?

13. Which point is located at $(0, 3)$?

14. Which point(s) are located in Quadrant IV?

15. Which point(s) are located in Quadrant III?

**Plot the point.**

16. $(3, -1)$

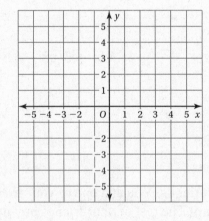

17. $(0, 2)$

18. $(-5, -4)$

19. $(-1, 0)$

20. $(-2, 3)$

Name_____ Date_____

# 13.1 Graphing Linear Equations
### For use with Exploration 13.1

**Learning Target:** Graph linear equations.

**Success Criteria:**
- I can create a table of values and write ordered pairs given a linear equation.
- I can plot ordered pairs to create a graph of a linear equation.
- I can use a graph of a linear equation to solve a real-life problem.

## 1 EXPLORATION: Creating Graphs

**Work with a partner. It starts snowing at midnight in Town A and Town B. The snow falls at a rate of 1.5 inches per hour.**

a. In Town A, there is no snow on the ground at midnight. How deep is the snow at each hour between midnight and 6 A.M.? Make a graph that represents this situation.

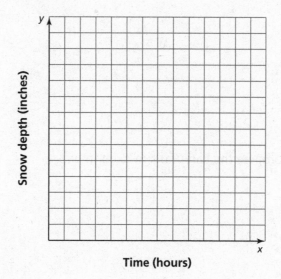

b. Repeat part (a) for Town B, which has 4 inches of snow on the ground at midnight.

**13.1** **Graphing Linear Equations** (continued)

**c.** The equations below represent the depth $y$ (in inches) of snow $x$ hours after midnight in Town C and Town D. Graph each equation.

**Town C**
$y = 2x + 3$

**Town D**
$y = 8$

**d.** Use your graphs to compare the snowfalls in each town.

## 13.1 Notetaking with Vocabulary

**Vocabulary:**

**Notes:**

## 13.1 Self-Assessment

Use the scale below to rate your understanding of the learning target and the success criteria.

| 1 | 2 | 3 | 4 |
|---|---|---|---|
| I do not understand. | I can do it with help. | I can do it on my own. | I can teach someone else. |

|  | Rating | Date |
|---|---|---|
| **13.1 Graphing Linear Equations** | | |
| **Learning Target:** Graph linear equations. | 1  2  3  4 | |
| I can create a table of values and write ordered pairs given a linear equation. | 1  2  3  4 | |
| I can plot ordered pairs to create a graph of a linear equation. | 1  2  3  4 | |
| I can use a graph of a linear equation to solve a real-life problem. | 1  2  3  4 | |

Name _____ Date _____

## 13.1 Practice

**Graph the linear equation.**

1. $y = 3.5$

2. $y = \frac{2}{3}x - 2$

3. $y = \frac{10}{3}x$

4. $y = -\frac{x}{2} + \frac{3}{2}$

5. The equation $y = 1.5x + 35$ represents the cost $y$ (in dollars) of the family meal when the food costs $35 and $x$ beverages are purchased.

   **a.** Graph the equation.

   **b.** Use the graph to estimate the cost of the family meal when 5 beverages are purchased.

   **c.** Use the equation to find the exact cost of the family meal when 5 beverages are purchased.

**Solve for $y$. Then graph the equation.**

6. $2y + 3x = -6$

7. $x + 0.25y = 1.5$

8. There are 10 coconuts at the base of your tree. The coconuts are falling off the tree at a rate of 6 coconuts per week. Assume that you do not pick up any coconuts.

   **a.** Write and graph a linear equation that represents the number of coconuts at the base of your tree after $x$ weeks.

   **b.** The tree will have no coconuts on it when there are 52 coconuts at the base of the tree. After how many weeks will this occur?

9. The sum $s$ of the first $n$ positive integers is $s = \frac{1}{2}n(n + 1)$.

   **a.** Plot four points $(n, s)$ that satisfy the equation. Is the equation a linear equation? Explain your reasoning.

   **b.** Does the value $n = 4.2$ make sense in the context of the problem? Explain your reasoning.

10. The equation $y = 5.50x$ represents the cost $y$ (in dollars) for $x$ visits to the science museum in a month. Graph the linear equation. What does the graph tell you about your purchase plan with the science museum?

## 13.2 Slope of a Line
### For use with Exploration 13.2

**Learning Target:** Find and interpret the slope of a line.

**Success Criteria:**
- I can explain the meaning of slope.
- I can find the slope of a line.
- I can interpret the slope of a line in a real-life problem.

### 1 EXPLORATION: Measuring the Steepness of a Line

**Work with a partner. Draw any nonvertical line in a coordinate plane.**

a. Develop a way to measure the *steepness* of the line. Compare your method with other pairs.

b. Draw a line that is parallel to your line. What can you determine about the steepness of each line? Explain your reasoning.

**13.2** **Slope of a Line** (continued)

**2** **EXPLORATION:** Using Right Triangles

**Work with a partner. Use the figure shown.**

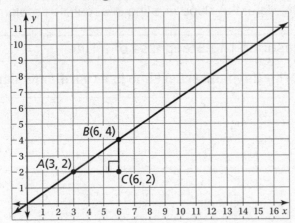

a. $\triangle ABC$ is a right triangle formed by drawing a horizontal line segment from point $A$ to a vertical line segment from point $B$. Use this method to draw another right triangle, $\triangle DEF$, with its longest side on the line.

b. What can you conclude about the two triangles in part (a)? Justify your conclusion. Compare your results with other pairs.

c. Based on your conclusions in part (b), what is true about $\dfrac{BC}{AC}$ and the corresponding measure in $\triangle DEF$? What do these values tell you about the line?

## 13.2 Notetaking with Vocabulary

**Vocabulary:**

**Notes:**

## 13.2 Self-Assessment

Use the scale below to rate your understanding of the learning target and the success criteria.

| 1 | 2 | 3 | 4 |
|---|---|---|---|
| I do not understand. | I can do it with help. | I can do it on my own. | I can teach someone else. |

| | Rating | Date |
|---|---|---|
| **13.2 Slope of a Line** | | |
| **Learning Target:** Find and interpret the slope of a line. | 1   2   3   4 | |
| I can explain the meaning of slope. | 1   2   3   4 | |
| I can find the slope of a line. | 1   2   3   4 | |
| I can interpret the slope of a line in a real-life problem. | 1   2   3   4 | |

Name _____ Date _____

## 13.2 Practice

**Find the slope of the line.**

1.

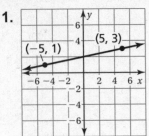

2.

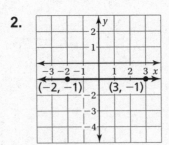

**The points in the table lie on a line. Find the slope of the line.**

3.

| x | 0 | 2 | 4 | 6 |
|---|---|---|---|---|
| y | -4 | -1 | 2 | 5 |

4.

| x | -4 | -1 | 0 | 3 |
|---|---|---|---|---|
| y | 7 | 4 | 3 | 0 |

5. A ramp used to remove furniture from a moving truck has a slope of $\frac{2}{5}$. The height of the ramp is 4 feet. How far does the base of the ramp extend from the end of the truck?

6. The graph shows the cost of a data usage on a phone plan.

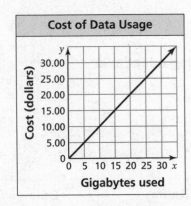

a. Find the slope of the line.

b. Explain the meaning of the slope as a rate of change.

c. How much money is added to the phone bill if you use 5 gigabytes of data?

d. How many gigabytes did you use if the data usage costs $30?

**Use an equation to find the value of k so that the line that passes through the given points has the given slope. Explain your reasoning.**

7. $(1, -1), (-2, k); m = -4$

8. $(-3, -4), (k, -2); m = \frac{2}{3}$

# 13.3 Graphing Proportional Relationships
### For use with Exploration 13.3

**Learning Target:** Graph proportional relationships.

**Success Criteria:**
- I can graph an equation that represents a proportional relationship.
- I can write an equation that represents a proportional relationship.
- I can use graphs to compare proportional relationships.

---

**1** **EXPLORATION:** Using a Ratio Table to Find Slope

**Work with a partner. The graph shows amounts of vinegar and water that can be used to make a cleaning product.**

   **a.** Use the graph to make a ratio table relating the quantities. Explain how the slope of the line is represented in the table.

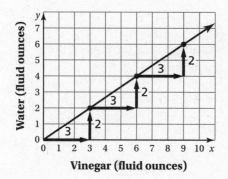

   **b.** Make a ratio table that represents a different ratio of vinegar to water. Use the table to describe the slope of the graph of the new relationship.

**13.3** Graphing Proportional Relationships (continued)

**2** **EXPLORATION:** Deriving an Equation

**Work with a partner. Let $(x, y)$ represent any point on the graph of a proportional relationship.**

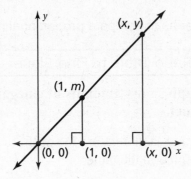

a. Describe the relationship between the corresponding side lengths of the triangles shown in the graph. Explain your reasoning.

b. Use the relationship in part (a) to write an equation relating $y$, $m$, and $x$. Then solve the equation for $y$.

c. What does your equation in part (b) describe? What does $m$ represent? Explain your reasoning.

**Notetaking with Vocabulary**

**Vocabulary:**

**Notes:**

## 13.3 Self-Assessment

Use the scale below to rate your understanding of the learning target and the success criteria.

| 1 | 2 | 3 | 4 |
|---|---|---|---|
| I do not understand. | I can do it with help. | I can do it on my own. | I can teach someone else. |

|  | Rating | Date |
|---|---|---|
| **13.3 Graphing Proportional Relationships** | | |
| **Learning Target:** Graph proportional relationships. | 1  2  3  4 | |
| I can graph an equation that represents a proportional relationship. | 1  2  3  4 | |
| I can write an equation that represents a proportional relationship. | 1  2  3  4 | |
| I can use graphs to compare proportional relationships. | 1  2  3  4 | |

Name _____ Date _____

# 13.3 Practice

**Tell whether *x* and *y* are in a proportional relationship. Explain your reasoning. If so, write an equation that represents the relationship.**

**1.**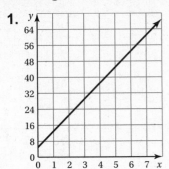

**2.**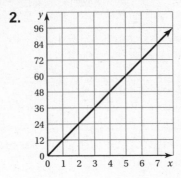

**3.**

| x | 2 | 5 | 8 | 11 |
|---|---|---|---|---|
| y | 8 | 20 | 32 | 44 |

**4.**

| x | 3 | 6 | 9 | 12 |
|---|---|---|---|---|
| y | 2 | 4 | 6 | 8 |

**5.** The cost *y* (in dollars) to rent a lane at bowling alley A is proportional to the number *x* of hours that you rent the lane. It costs $18 to rent the lane for 2 hours.

   **a.** Write an equation that represents the situation.

   **b.** Interpret the slope of the graph of the equation.

   **c.** How much does it cost to rent the lane for 3 hours?

   **d.** At bowling alley B it costs $16.50 to rent a lane for 2 hours. Write an equation that represents the situation.

   **e.** In the same coordinate plane, graph equations that represent the costs of renting a lane for 2 hours at bowling alley A and bowling alley B. Compare and interpret the steepness of each graph.

**6.** The graph relates the height of the water in a tank *y* (in inches) to the volume of the water *x* (in gallons).

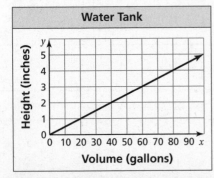

   **a.** Is the relationship proportional? Explain.

   **b.** Write an equation of the line. Interpret the slope.

   **c.** What is the height of the water in the tank when the volume is 250 gallons?

## 13.4 Graphing Linear Equations in Slope-Intercept Form
**For use with Exploration 13.4**

**Learning Target:** Graph linear equations in slope-intercept form.

**Success Criteria:** • I can identify the slope and *y*-intercept of a line given an equation.
• I can rewrite a linear equation in slope-intercept form.
• I can use the slope and *y*-intercept to graph linear equations.

---

**1 EXPLORATION: Deriving an Equation**

**Work with a partner. In the previous section, you learned that the graph of a proportional relationship can be represented by the equation $y = mx$, where $m$ is the constant of proportionality.**

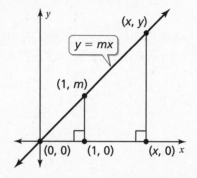

a. You translate the graph of a proportional relationship 3 units up as shown below. Let $(x, y)$ represent any point on the graph. Make a conjecture about the equation of the line. Explain your reasoning.

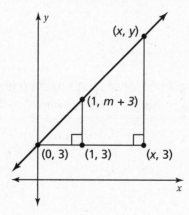

---

**13.4** **Graphing Linear Equations in Slope-Intercept Form** (continued)

**b.** Describe the relationship between the corresponding side lengths of the triangles. Explain your reasoning.

**c.** Use the relationship in part (b) to write an equation relating $y$, $m$, and $x$. Does your equation support your conjecture in part (a)? Explain.

**d.** You translate the graph of a proportional relationship $b$ units up. Write an equation relating $y$, $m$, $x$, and $b$. Justify your answer.

## 13.4 Notetaking with Vocabulary

**Vocabulary:**

**Notes:**

## 13.4 Self-Assessment

**Use the scale below to rate your understanding of the learning target and the success criteria.**

| 1 | 2 | 3 | 4 |
|---|---|---|---|
| I do not understand. | I can do it with help. | I can do it on my own. | I can teach someone else. |

|  | Rating | Date |
|---|---|---|
| **13.4 Graphing Linear Equations in Slope-Intercept Form** | | |
| **Learning Target:** Graph linear equations in slope-intercept form. | 1   2   3   4 | |
| I can identify the slope and y-intercept of a line given an equation. | 1   2   3   4 | |
| I can rewrite a linear equation in slope-intercept form. | 1   2   3   4 | |
| I can use the slope and y-intercept to graph linear equations. | 1   2   3   4 | |

Name _____ Date _____

## 13.4 Practice

**Find the slope and the *y*-intercept of the graph of the linear equation.**

**1.** $y = -\frac{3}{8}x + 10$

**2.** $y + \frac{1}{5} = -\frac{4}{5}x$

**3.** The number $y$ of gallons of water in the swimming pool $x$ minutes after turning on the faucet is represented by $y = 24x + 285$.

    **a.** Graph the linear equation.

    **b.** Interpret the slope and the *y*-intercept.

    **c.** Is the *x*-intercept applicable to the problem? Explain.

**Graph the linear equation. Identify the *x*-intercept.**

**4.** $y = -1.2x + 9$

**5.** $y + 3 = -\frac{6}{7}x$

**6.** There is a $10 monthly membership fee to download music. There is a $0.50 fee for each song downloaded.

    **a.** Write a linear equation that models the cost of downloading $x$ songs per month.

    **b.** Graph the equation.

    **c.** What is the cost of downloading 15 songs? Explain your reasoning.

**7.** An entrepreneur is opening a business to market pies and pie fillings based on her family's recipes. The price of every item in the store is $6.

    **a.** Write a linear equation that models the amount of revenue $y$ (in dollars) taken in for selling $x$ items.

    **b.** Graph the equation.

    **c.** The monthly cost of rent and utilities for the store space is $1100. What is the minimum number of items that must be sold each month in order to make a profit? Explain your reasoning.

    **d.** Assuming 4 weeks in a month, what is the average number of items that need to be sold each week in order to turn a profit?

# 13.5 Graphing Linear Equations in Standard Form
**For use with Exploration 13.5**

**Learning Target:** Graph linear equations in standard form.

**Success Criteria:**
- I can rewrite the standard form of a linear equation in slope-intercept form.
- I can find intercepts of linear equations written in standard form.
- I can use intercepts to graph linear equations.

## 1 EXPLORATION: Using Intercepts

**Work with a partner. You spend $150 on fruit trays and vegetable trays for a party.**

**Fruit Tray: $50**                    **Vegetable Tray: $25**

a. You buy *x* fruit trays and *y* vegetable trays. Complete the verbal model. Then use the verbal model to write an equation that relates *x* and *y*.

$$\frac{\underline{\quad\quad}}{1 \text{ fruit tray}} \cdot \begin{array}{c}\text{Number}\\ \text{of fruit}\\ \text{trays}\end{array} + \frac{\underline{\quad\quad}}{1 \text{ vegetable tray}} \cdot \begin{array}{c}\text{Number}\\ \text{of vegetable}\\ \text{trays}\end{array} = \underline{\quad}$$

**13.5** **Graphing Linear Equations in Standard Form** (continued)

**b.** What is the greatest number of fruit trays that you can buy? vegetable trays? Can you use these numbers to graph your equation from part (a) in the coordinate plane? Explain.

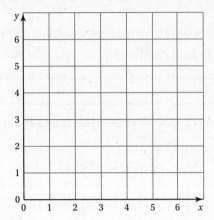

**c.** Use a graph to determine the different combinations of fruit trays and vegetable trays that you can buy. Justify your answers algebraically.

**d.** You are given an extra $50 to spend. How does this affect the intercepts of your graph in part (c)? Explain your reasoning?

## 13.5 Notetaking with Vocabulary

**Vocabulary:**

**Notes:**

## 13.5 Self-Assessment

Use the scale below to rate your understanding of the learning target and the success criteria.

| **1** | **2** | **3** | **4** |
|---|---|---|---|
| I do not understand. | I can do it with help. | I can do it on my own. | I can teach someone else. |

| | Rating | Date |
|---|---|---|
| **13.5 Graphing Linear Equations in Standard Form** | | |
| **Learning Target:** Graph linear equations in standard form. | 1   2   3   4 | |
| I can rewrite the standard form of a linear equation in slope-intercept form. | 1   2   3   4 | |
| I can find intercepts of linear equations written in standard form. | 1   2   3   4 | |
| I can use intercepts to graph linear equations. | 1   2   3   4 | |

## 13.5 Practice

**Write the linear equation in slope-intercept form.**

**1.** $\frac{2}{3}x + y = 4$

**2.** $4x - 2y = 10$

**Graph the linear equation.**

**3.** $4.5x - 0.5y = 3$

**4.** $\frac{2}{3}x + \frac{1}{3}y = 2$

**5.** You work at the local pool as a lifeguard and you also work in the snack bar. You earn $15 per hour lifeguarding and $8 per hour working in the snack bar. Last week you worked a total of 20 hours and earned $202.

   **a.** Write an equation in standard form for the hours you worked.

   **b.** Write an equation in standard form for the money you earned.

   **c.** Graph both equations on the same coordinate plane.

   **d.** Determine how many hours you worked as a lifeguard and how many hours you worked in the snack bar. Explain your reasoning.

**Graph the linear equation using intercepts.**

**6.** $\frac{1}{5}x + \frac{1}{10}y = \frac{2}{5}$

**7.** $2.5x - 1.25y = 5$

**8.** Your family is on a ski vacation. Lift tickets for the family cost $80 per day. Snowboard rentals cost $40 per day. You purchase lift tickets for $x$ days and snowboard rentals for $y$ days and spend $480.

   **a.** Write an equation in standard form that represents the situation.

   **b.** Find the $x$- and $y$-intercepts.

   **c.** Graph the equation.

   **d.** You rent snowboards for 2 days. How many days did you purchase lift tickets?

**9.** An electrician charges $80 plus $32 per hour.

   **a.** Write an equation that represents the total fee $y$ (in dollars) charged by the electrician for a job lasting $x$ hours.

   **b.** Find the $x$- and $y$-intercepts.

   **c.** Graph the equation.

   **d.** Is the value of the $x$-intercept applicable to the electrician? Explain.

# 13.6 Writing Equations in Slope-Intercept Form
**For use with Exploration 13.6**

**Learning Target:** Write equations of lines in slope-intercept form.

**Success Criteria:**
- I can find the slope and the *y*-intercept of a line.
- I can use the slope and the *y*-intercept to write an equation of a line.
- I can write equations in slope-intercept form to solve real-life problems.

---

**1  EXPLORATION:** Writing Equations of Lines

**Work with a partner. For each part, answer the following questions.**

- **What are the slopes and the *y*-intercepts of the lines?**

- **What are equations that represent the lines?**

- **What do the lines have in common?**

**a.**

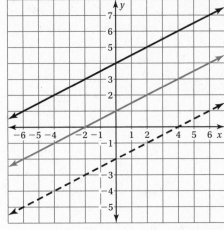

**b.**

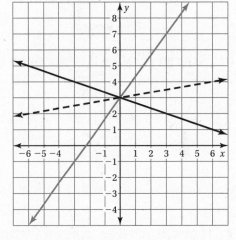

---

**13.6** **Writing Equations in Slope-Intercept Form** (continued)

**2** **EXPLORATION:** Interpreting the Slope and the *y*-intercept

**Work with a partner. The graph represents the distance *y* (in miles) of a car from Phoenix after *t* hours of a trip.**

   **a.** Find the slope and the *y*-intercept of the line. What do they represent in this situation?

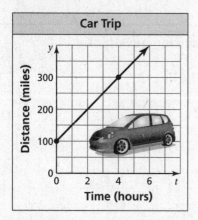

   **b.** Write an equation that represents the graph.

   **c.** How can you determine the distance of the car from Phoenix after 11 hours?

## 13.6 Notetaking with Vocabulary

**Vocabulary:**

**Notes:**

## 13.6 Self-Assessment

Use the scale below to rate your understanding of the learning target and the success criteria.

| *1* | *2* | *3* | *4* |
|---|---|---|---|
| I do not understand. | I can do it with help. | I can do it on my own. | I can teach someone else. |

|  | Rating | Date |
|---|---|---|
| **13.6 Writing Equations in Slope-Intercept Form** | | |
| **Learning Target:** Write equations of lines in slope-intercept form. | 1   2   3   4 | |
| I can find the slope and the y-intercept of a line. | 1   2   3   4 | |
| I can use the slope and the y-intercept to write an equation of a line. | 1   2   3   4 | |
| I can write equations in slope-intercept form to solve real-life problems. | 1   2   3   4 | |

# 13.6 Practice

1. Write an equation that represents each side of the figure.

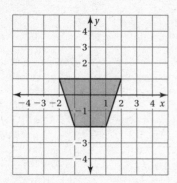

**Write an equation in slope-intercept form of the line that passes through the given points.**

2.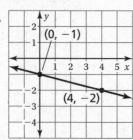

3.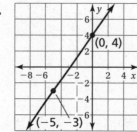

4. Your hair is 6 inches long and grows at a rate of 144 millimeters per year.

   a. Convert 144 millimeters per year to inches per year. Round your answer to the nearest tenth.

   b. Write an equation that represents the length $y$ (in inches) of your hair after $x$ years.

   c. If you do not cut it, how long is your hair after 4 years?

**Write an equation of the line that passes through the given points.**

5. $(-4, -1), (0, 5)$

6. $(0, -3), (1, -5)$

7. Yesterday, you typed 8 pages in 48 minutes. Today, you typed 20 pages in 2 hours.

   a. Plot the two points $(x, y)$, where $x$ is the time (in minutes) and $y$ is the number of pages.

   b. What is the rate of typing? Explain.

   c. Write an equation that represents the number of pages in terms of the number of minutes.

## 13.7 Writing Equations in Point-Slope Form
**For use with Exploration 13.7**

**Learning Target:** Write equations of lines in point-slope form.

**Success Criteria:**
- I can use a point on a line and the slope to write an equation of the line.
- I can use any two points to write an equation of a line.
- I can write equations in point-slope form to solve real-life problems.

---

**1 EXPLORATION:** Deriving an Equation

**Work with a partner. Let $(x_1, y_1)$ represent a specific point on a line. Let $(x, y)$ represent any other point on the line.**

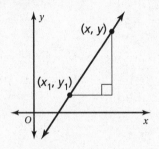

a. Write an equation that represents the slope $m$ of the line. Explain your reasoning.

b. Multiply each side of your equation in part (a) by the expression in the denominator. What does the resulting equation represent? Explain your reasoning.

## 13.7 Writing Equations in Point-Slope Form (continued)

**2  EXPLORATION: Writing an Equation**

**Work with a partner.**

For 4 months, you saved $25 a month. You now have $175 in your savings account.

**a.** Draw a graph that shows the balance in your account after $t$ months.

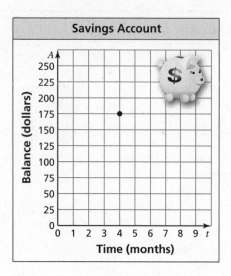

**b.** Use your result from Exploration 1 to write an equation that represents the balance $A$ after $t$ months.

Name_____ Date_____

 **13.7** **Notetaking with Vocabulary**

**Vocabulary:**

**Notes:**

**13.7** **Self-Assessment**

Use the scale below to rate your understanding of the learning target and the success criteria.

| 1 | 2 | 3 | 4 |
|---|---|---|---|
| I do not understand. | I can do it with help. | I can do it on my own. | I can teach someone else. |

|  | Rating | Date |
|---|---|---|
| **13.7 Writing Equations in Point-Slope Form** | | |
| **Learning Target:** Write equations of lines in point-slope form. | 1  2  3  4 | |
| I can use a point on a line and the slope to write an equation of the line. | 1  2  3  4 | |
| I can use any two points to write an equation of a line. | 1  2  3  4 | |
| I can write equations in point-slope form to solve real-life problems. | 1  2  3  4 | |

## 13.7 Practice

**Write an equation of the line with the given slope that passes through the given point. Graph the line.**

**1.** $m = \dfrac{5}{4}$

**2.** $m = -4$

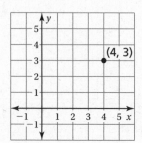

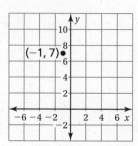

**Write an equation in point-slope form of the line that passes through the given point and has the given slope.**

**3.** $(-6, 3); m = \dfrac{1}{3}$

**4.** $(8, -7); m = -\dfrac{3}{4}$

**Write an equation in slope-intercept form of the line that passes through the given points.**

**5.** $(2, 3), (3, 7)$

**6.** $(-5, -8), (10, 4)$

**7.** The cost of renting the party room for 10 people is $117.50. The cost of renting the room is $151.25 for 15 people. Write an equation that represents the cost of renting the party room in terms of the number of people attending the party.

**8.** You are pulling a kite back to the ground at a rate of 2 feet per second. After 4 seconds, the kite is 16 feet above the ground.

   **a.** Write an equation that represents the height $y$ (in feet) of the kite above the ground after $x$ seconds.

   **b.** At what height was the kite when you started pulling it in?

   **c.** When does the kite touch the ground?

**9.** Write an equation of the line that passes through the point (-12, 6) and is parallel to the graph of the equation $y = -\dfrac{1}{6}x + 1$. Explain your reasoning.

## Chapter 13 Chapter Self-Assessment

**Use the scale below to rate your understanding of the learning target and the success criteria.**

| 1 | 2 | 3 | 4 |
|---|---|---|---|
| I do not understand. | I can do it with help. | I can do it on my own. | I can teach someone else. |

|  | Rating | Date |
|---|---|---|
| **13.1 Graphing Linear Equations** | | |
| **Learning Target:** Graph linear equations. | 1  2  3  4 | |
| I can create a table of values and write ordered pairs given a linear equation. | 1  2  3  4 | |
| I can plot ordered pairs to create a graph of a linear equation. | 1  2  3  4 | |
| I can use a graph of a linear equation to solve a real-life problem. | 1  2  3  4 | |
| **13.2 Slope of a Line** | | |
| **Learning Target:** Find and interpret the slope of a line. | 1  2  3  4 | |
| I can explain the meaning of slope. | 1  2  3  4 | |
| I can find the slope of a line. | 1  2  3  4 | |
| I can interpret the slope of a line in a real-life problem. | 1  2  3  4 | |
| **13.3 Graphing Proportional Relationships** | | |
| **Learning Target:** Graph proportional relationships. | 1  2  3  4 | |
| I can graph an equation that represents a proportional relationship. | 1  2  3  4 | |
| I can write an equation that represents a proportional relationship. | 1  2  3  4 | |
| I can use graphs to compare proportional relationships. | 1  2  3  4 | |

Name _____ Date _____

| | Rating | Date |
|---|---|---|
| **13.4 Graphing Linear Equations in Slope-Intercept Form** | | |
| **Learning Target:** Graph linear equations in slope-intercept form. | 1   2   3   4 | |
| I can identify the slope and $y$-intercept of a line given an equation. | 1   2   3   4 | |
| I can rewrite a linear equation in slope-intercept form. | 1   2   3   4 | |
| I can use the slope and $y$-intercept to graph linear equations. | 1   2   3   4 | |
| **13.5 Graphing Linear Equations in Standard Form** | | |
| **Learning Target:** Graph linear equations in standard form. | 1   2   3   4 | |
| I can rewrite the standard form of a linear equation in slope-intercept form. | 1   2   3   4 | |
| I can find intercepts of linear equations written in standard form. | 1   2   3   4 | |
| I can use intercepts to graph linear equations. | 1   2   3   4 | |
| **13.6 Writing Equations in Slope-Intercept Form** | | |
| **Learning Target:** Write equations of lines in slope-intercept form. | 1   2   3   4 | |
| I can find the slope and the $y$-intercept of a line. | 1   2   3   4 | |
| I can use the slope and the $y$-intercept to write an equation of a line. | 1   2   3   4 | |
| I can write equations in slope-intercept form to solve real-life problems. | 1   2   3   4 | |
| **13.7 Writing Equations in Point-Slope Form** | | |
| **Learning Target:** Write equations of lines in point-slope form. | 1   2   3   4 | |
| I can use a point on a line and the slope to write an equation of the line. | 1   2   3   4 | |
| I can use any two points to write an equation of a line. | 1   2   3   4 | |
| I can write equations in point-slope form to solve real-life problems. | 1   2   3   4 | |

## Chapter 14   Review & Refresh

**Evaluate the expression.**

**1.** $2 + 1 \cdot 4^2 - 12 \div 3$

**2.** $8^2 \div 16 \cdot 2 - 5$

**3.** $7(9 - 3) + 6^2 \cdot 10 - 8$

**4.** $3 \cdot 5 - 10 + 9(2 + 1)^2$

**5.** $8(6 + 5) - (9^2 + 3) \div 7$

**6.** $5[3(12 - 8)] - 6 \cdot 8 + 2^2$

**7.** $4 + 4 + 5 \times 2 \times 5 + (3 + 3 + 3) \times 6 \times 6 + 2 + 2$

   **a.** Evaluate the expression.

   **b.** Rewrite the expression using what you know about order of operations. Then evaluate.

**Chapter 14** **Review & Refresh** (continued)

**Find the product or quotient.**

**8.** $3.92 \cdot 0.6$                         **9.** $0.78 \cdot 0.13$

**10.** $5.004 \times 1.2$                  **11.** $6.3 \div 0.7$

**12.** $2.25 \div 1.5$                   **13.** $8.1 \div 0.003$

**14.** Grapes cost $1.98 per pound. You buy 3.5 pounds of grapes. How much do you pay for the grapes?

## 14.1 Exponents
**For use with Exploration 14.1**

**Learning Target:** Use exponents to write and evaluate expressions.

**Success Criteria:**
- I can write products using exponents.
- I can evaluate expressions involving powers.
- I can use exponents to solve real-life problems.

---

**1  EXPLORATION:** Using Exponent Notation

**Work with a partner.**

a. Complete the table.

| Power | Repeated Multiplication Form | Value |
|---|---|---|
| $(-3)^1$ | $-3$ | $-3$ |
| $(-3)^2$ | $(-3) \cdot (-3)$ | $9$ |
| $(-3)^3$ | | |
| $(-3)^4$ | | |
| $(-3)^5$ | | |
| $(-3)^6$ | | |
| $(-3)^7$ | | |

b. Describe what is meant by the expression $(-3)^n$. How can you find the value of $(-3)^n$?

**14.1** **Exponents** (continued)

---

**2** **EXPLORATION:** Using Exponent Notation

**Work with a partner. On a game show, each small cube is worth $3. The small cubes are arranged to form a large cube with a side length of three small cubes. Show how you can use a power to find the total value of the large cube. Then write an explanation to convince a friend that your answer is correct.**

---

## 14.1 Notetaking with Vocabulary

**Vocabulary:**

**Notes:**

## 14.1 Self-Assessment

Use the scale below to rate your understanding of the learning target and the success criteria.

| **1** | **2** | **3** | **4** |
|---|---|---|---|
| I do not understand. | I can do it with help. | I can do it on my own. | I can teach someone else. |

| | Rating | Date |
|---|---|---|
| **14.1 Exponents** | | |
| **Learning Target:** Use exponents to write and evaluate expressions. | 1  2  3  4 | |
| I can write products using exponents. | 1  2  3  4 | |
| I can evaluate expressions involving powers. | 1  2  3  4 | |
| I can use exponents to solve real-life problems. | 1  2  3  4 | |

## 14.1 Practice

**Write the product using exponents.**

1. $-\dfrac{3}{7} \cdot \dfrac{3}{7} \cdot \dfrac{3}{7}$

2. $\left(-\dfrac{3}{7}\right) \cdot \left(-\dfrac{3}{7}\right) \cdot \left(-\dfrac{3}{7}\right)$

3. $25 \cdot 25 \cdot 25 \cdot 25 \cdot (-p) \cdot (-p) \cdot (-p) \cdot (-p) \cdot (-p)$

4. $(-2) \cdot (-2) \cdot x \cdot x \cdot x \cdot y \cdot y \cdot y \cdot y$

**Evaluate the expression.**

5. $7^3$

6. $-4^4$

7. $(-4)^4$

8. $\left(\dfrac{2}{5}\right)^3$

9. Write the prime factorization of 1323 using exponents.

**Evaluate the expression.**

10. $280 - (-3) \cdot (-5)^3$

11. $(20^2 - 3^3 \cdot 8^2) \div 16$

12. $\dfrac{2}{3}(16^2 - 17^2)$

13. $\left| \dfrac{1}{5}\left(\dfrac{6^3}{3^3} - 2^3\right) \right|$

14. Bed A is 7 feet long. Bed B is $\dfrac{7}{8}$ as long as bed A. Bed C is $\dfrac{7}{8}$ as long as Bed B. Bed D is $\dfrac{7}{8}$ as long as bed C.

    a. Write an expression for the length of Bed D.

    b. What is the length of Bed D?

15. Complete the table. Compare the values of $3^h - 2$ with the values of $3^{h-1}$. When are the values the same?

| $h$ | 1 | 2 | 3 | 4 | 5 |
|---|---|---|---|---|---|
| $3^h - 2$ | | | | | |
| $3^{h-1}$ | | | | | |

16. You ran 8 miles. John ran half as far as you. Tim ran half as far as John. Chris ran half as far as Tim.

    a. Write an expression for how far Chris ran.

    b. How far did Chris run?

## 14.2 Product of Powers Property
**For use with Exploration 14.2**

**Learning Target:** Generate equivalent expressions involving products of powers.

**Success Criteria:**
- I can find products of powers that have the same base.
- I can find powers of powers.
- I can find powers of products.

---

**1** **EXPLORATION: Finding Products of Powers**

**Work with a partner.**

**a.** Complete the table. Use your results to write a *general rule* for finding $a^m \cdot a^n$, a product of two powers with the same base.

| Product | Repeated Multiplication Form | Power |
|---|---|---|
| $2^2 \cdot 2^4$ | | |
| $(-3)^2 \cdot (-3)^4$ | | |
| $7^3 \cdot 7^2$ | | |
| $5.1^1 \cdot 5.1^6$ | | |
| $(-4)^2 \cdot (-4)^2$ | | |
| $10^3 \cdot 10^5$ | | |
| $\left(\frac{1}{2}\right)^5 \cdot \left(\frac{1}{2}\right)^5$ | | |

**b.** Show how to use your rule in part (a) to write each expression below as a single power. Then write a *general rule* for finding $(a^m)^n$, a power of a power.

$(7^3)^2$ $\qquad$ $(6^2)^2$ $\qquad$ $(3^2)^3$ $\qquad$ $(2^2)^4$ $\qquad$ $\left(\left(\frac{1}{2}\right)^2\right)^5$

**14.2** **Product of Powers Property** (continued)

---

**2** **EXPLORATION:** Finding Powers of Products

**Work with a partner. Complete the table. Use your results to write a *general rule* for finding $(ab)^m$, a power of a product.**

| Product | Repeated Multiplication Form | Power |
|---|---|---|
| $(2 \cdot 3)^3$ | | |
| $(2 \cdot 5)^2$ | | |
| $(5 \cdot 4)^3$ | | |
| $(-2 \cdot 4)^2$ | | |
| $(-3 \cdot 2)^4$ | | |

# 14.2 Notetaking with Vocabulary

**Vocabulary:**

**Notes:**

# 14.2 Self-Assessment

Use the scale below to rate your understanding of the learning target and the success criteria.

| *1* | *2* | *3* | *4* |
|---|---|---|---|
| I do not understand. | I can do it with help. | I can do it on my own. | I can teach someone else. |

|  | Rating | Date |
|---|---|---|
| **14.2 Product of Powers Property** | | |
| **Learning Target:** Generate equivalent expressions involving products of powers. | 1　2　3　4 | |
| I can find products of powers that have the same base. | 1　2　3　4 | |
| I can find powers of powers. | 1　2　3　4 | |
| I can find powers of products. | 1　2　3　4 | |

## 14.2 Practice

**Simplify the expression. Write your answer as a power.**

**1.** $(-16)^5 \cdot (-16)^{21}$

**2.** $\left(\frac{1}{15}\right)^{12} \cdot \left(\frac{1}{15}\right)$

**3.** $q^7 \cdot q^9$

**4.** $(-7.4)^9 \cdot (-7.4)^{12}$

**5.** $\left(\left(\frac{5}{8}\right)^2\right)^3$

**6.** $\left(\left(-\frac{2}{9}\right)^3\right)^5$

**Simplify the expression.**

**7.** $(-2p)^4$

**8.** $\left(\frac{1}{5}k\right)^3$

**9.** $(3^2)^4 - 3^5 \cdot 3$

**10.** $10\left(\frac{1}{5}v\right)^3$

**11.** In 2016 about $(2 \cdot 5)^4 \cdot 3^3$ text messages were sent every second. There are $2^7(3 \cdot 5)^2$ seconds in 1 day. How many text messages were sent each day in 2016? Write your answer as an expression involving three powers and in standard form.

**12.** The volume of a right circular cylinder is $V = \pi r^2 h$. The relationship between the height $h$ of a given right circular cylinder and the radius $r$ is $r = \frac{2}{3}h$.

  **a.** Find the volume of the right circular cylinder in terms of the height $h$ and simplify the expression.

  **b.** What is the volume of the right circular cylinder when the height is $\frac{3}{4}$ inch?

  **c.** Find the volume of the right circular cylinder in terms of the radius $r$ and simplify the expression.

**13.** Show that $(5 \cdot 27 \cdot y)^6 = 15^6 \cdot 9^6 \cdot y^6$.

**Find the value of *x* in the equation without evaluating the power. Explain your reasoning.**

**14.** $3^2 \cdot 3^x = 3^{12}$

**15.** $(5^x)^4 = 5^{24}$

**16.** Your boss tells you that you have to work every day in February (28 days). He gives you the choice to be paid $1000 per day or $2^x$ cents per day, where $x$ is the day of the month. Which payment method do you choose? What is the total amount of money you get paid at the end of the month?

## 14.3 Quotient of Powers Property
**For use with Exploration 14.3**

**Learning Target:** Generate equivalent expressions involving quotients of powers.

**Success Criteria:**
- I can find quotients of powers that have the same base.
- I can simplify expressions using the Quotient of Powers Property.
- I can solve real-life problems involving quotients of powers.

---

**1 EXPLORATION:** Finding Quotients of Powers

**Work with a partner.**

**a.** Complete the table. Use your results to write a *general rule* for finding $\dfrac{a^m}{a^n}$, a quotient of two powers with the same base.

| Product | Repeated Multiplication Form | Power |
|---|---|---|
| $\dfrac{2^4}{2^2}$ | $\dfrac{2 \cdot 2 \cdot 2 \cdot 2}{2 \cdot 2}$ | |
| $\dfrac{(-4)^5}{(-4)^2}$ | | |
| $\dfrac{7^7}{7^3}$ | | |
| $\dfrac{8.5^9}{8.5^6}$ | | |
| $\dfrac{10^8}{10^5}$ | | |
| $\dfrac{3^{12}}{3^4}$ | | |
| $\dfrac{(-5)^7}{(-5)^5}$ | | |
| $\dfrac{11^4}{11^1}$ | | |
| $\dfrac{x^6}{x^2}$ | | |

**14.3** **Quotient of Powers Property** (continued)

**b.** Use your rule in part (a) to simplify the quotients in the first column of the table above. Does your rule give the results in the third column?

**Notetaking with Vocabulary**

**Vocabulary:**

**Notes:**

 **Self-Assessment**

Use the scale below to rate your understanding of the learning target and the success criteria.

| 1 | 2 | 3 | 4 |
|---|---|---|---|
| I do not understand. | I can do it with help. | I can do it on my own. | I can teach someone else. |

| | Rating | Date |
|---|---|---|
| **14.3 Quotient of Powers Property** | | |
| **Learning Target:** Generate equivalent expressions involving quotients of powers. | 1  2  3  4 | |
| I can find quotients of powers that have the same base. | 1  2  3  4 | |
| I can simplify expressions using the Quotient of Powers Property. | 1  2  3  4 | |
| I can solve real-life problems involving quotients of powers. | 1  2  3  4 | |

Name _____  Date _____

## 14.3 Practice

**Simplify the expression. Write your answer as a power.**

1. $\dfrac{7.6^{13}}{7.6^3}$

2. $\dfrac{u^{33}}{u^{11}}$

3. One kilometer equals $10^3$ meters. One tetrameter equals $10^{12}$ meters.

   a. How many times larger is a tetrameter than a kilometer?

   b. A square has side length of 1 kilometer. Find the area in meters. Write your answer as a power.

   c. A square has side length of 1 tetrameter. Find the area in meters. Write your answer as a power.

   d. How many times larger is the area of the square in part (c) than the area of the square in part (b)?

   e. A cube has side length of 1 kilometer. Find the volume in meters. Write your answer as a power.

   f. A cube has side length of 1 tetrameter. Find the volume in meters. Write your answer as a power.

   g. How many times larger is the volume of the cube in part (e) than the volume of the cube in part (f)?

**Simplify the expression. Write your answer as a power.**

4. $\dfrac{(-7.9)^{15} \cdot (-7.9)^9}{(-7.9)^{12} \cdot (-7.9)^7}$

5. $\dfrac{b^{35}}{b^{20}} \cdot \dfrac{b^{15}}{b^{10}}$

**Determine whether the statement is *always*, *never*, or *sometimes* true. Explain your reasoning.**

6. The expression $\dfrac{5^{x+3}}{5^x}$ will equal $5^3$.

7. The expression $\dfrac{5^x}{5^y}$ will equal $5^3$.

8. The expression $\dfrac{5^{x+1}}{5^x}$ will equal $5^3$.

**Find the value of $x$ in the equation without evaluating the power.**

9. $\dfrac{9^7}{9^x} = 729$

10. $\dfrac{2^{12} \cdot 2^x}{2^{10}} = 32$

## 14.4 Zero and Negative Exponents
**For use with Exploration 14.4**

**Learning Target:** Understand the concepts of zero and negative exponents.

**Success Criteria:**
- I can explain the meanings of zero and negative exponents.
- I can evaluate numerical expressions involving zero and negative exponents.
- I can simplify algebraic expressions involving zero and negative exponents.

### 1 EXPLORATION: Understanding Zero Exponents

**Work with a partner.**

a. Complete the table.

| Quotient | Quotient of Powers Property | Power |
|---|---|---|
| $\dfrac{5^3}{5^3}$ | | |
| $\dfrac{6^2}{6^2}$ | | |
| $\dfrac{(-3)^4}{(-3)^4}$ | | |
| $\dfrac{(-4)^5}{(-4)^5}$ | | |

b. Evaluate each expression in the first column of the table in part (a). How can you use these results to define $a^0$, where $a \neq 0$?

**14.4** Zero and Negative Exponents (continued)

**2** **EXPLORATION:** Understanding Negative Exponents

**Work with a partner.**

a. Complete the table.

| Product | Product of Powers Property | Power | Value |
|---|---|---|---|
| $5^{-3} \cdot 5^3$ | | | |
| $6^2 \cdot 6^{-2}$ | | | |
| $(-3)^4 \cdot (-3)^{-4}$ | | | |
| $(-4)^{-5} \cdot (-4)^5$ | | | |

b. How can you use the Multiplicative Inverse Property to rewrite the powers containing negative exponents in the first column of the table?

c. Use your results in parts (a) and (b) to define $a^{-n}$, where $a \neq 0$ and $n$ is an integer.

## 14.4 Notetaking with Vocabulary

**Vocabulary:**

**Notes:**

## 14.4 Self-Assessment

Use the scale below to rate your understanding of the learning target and the success criteria.

| 1 | 2 | 3 | 4 |
|---|---|---|---|
| I do not understand. | I can do it with help. | I can do it on my own. | I can teach someone else. |

| | Rating | Date |
|---|:---:|:---:|
| **14.4 Zero and Negative Exponents** | | |
| **Learning Target:** Understand the concepts of zero and negative exponents. | 1   2   3   4 | |
| I can explain the meanings of zero and negative exponents. | 1   2   3   4 | |
| I can evaluate numerical expressions involving zero and negative exponents. | 1   2   3   4 | |
| I can simplify algebraic expressions involving zero and negative exponents. | 1   2   3   4 | |

## 14.4 Practice

**Evaluate the expression.**

1. $10^{-1} \cdot 10^{-2}$

2. $\dfrac{1}{3^{-4}} \cdot \dfrac{1}{3^6}$

3. $27^{-18} \cdot 27^{18}$

4. $\dfrac{4^{-7}}{4^2 \cdot 4^{-5}}$

5. Write three different powers with negative exponents that are equal to $64^{-1}$.

6. One millimeter equals $10^{-3}$ meter. One picometer equals $10^{-12}$ meter. One femtometer equals $10^{-15}$ meter.

   a. Find the product of one millimeter and one picometer, using only positive exponents.

   b. Find the quotient of one picometer and one millimeter, using only positive exponents.

   c. Find the product of one millimeter and one femtometer, using only positive exponents.

   d. Find the quotient of one femtometer and one picometer, using only positive exponents.

   e. Find the quotient of one picometer and one femtometer, using only positive exponents.

   f. Find the quotient of one millimeter and one femtometer, using only positive exponents.

   g. Find the product of one picometer and one femtometer, using only positive exponents.

**Simplify. Write the expression using only positive exponents.**

7. $\dfrac{14u^{-4}}{7u^8}$

8. $\dfrac{2^{-3} \cdot a^0 \cdot b^5}{b^{-4}}$

9. A swimming pool contains $6.8 \times 10^3$ gallons of water. The pool is leaking at a rate of $12^{-1}$ gallons per second. How many hours will it take for all of the water to leak out of the pool?

Name_____ Date_____

## 14.5 Estimating Quantities
### For use with Exploration 14.5

**Learning Target:** Round numbers and write the results as the product of a single digit and a power of 10.

**Success Criteria:**
- I can round very large and very small numbers.
- I can write a multiple of 10 as a power.
- I can compare very large or very small quantities.

---

**1  EXPLORATION:** Using Powers of 10

**Work with a partner. Match each picture with the most appropriate distance. Explain your reasoning.**

$6 \times 10^3$ m       $1 \times 10^1$ m       $2 \times 10^{-1}$ m       $6 \times 10^{-2}$ m

a.

b.

c.

d.

**14.5** **Estimating Quantities** (continued)

**2** **EXPLORATION:** Approximating Numbers

**Work with a partner. Match each number in List 1 with its closest approximation in List 2. Explain your method.**

| | *List 1* | | | *List 2* |
|---|---|---|---|---|
| **a.** | 180,000,000,000,000 | | **A.** | $3 \times 10^{11}$ |
| **b.** | 0.0000000011 | | **B.** | $1 \times 10^{-5}$ |
| **c.** | 302,000,000,000 | | **C.** | $2 \times 10^{14}$ |
| **d.** | 0.00000028 | | **D.** | $3 \times 10^{13}$ |
| **e.** | 0.0000097 | | **E.** | $3 \times 10^{-7}$ |
| **f.** | 330,000,000,000,000 | | **F.** | $1 \times 10^{-9}$ |
| **g.** | 26,000,000,000,000 | | **G.** | $2 \times 10^{-5}$ |
| **h.** | 0.000023 | | **H.** | $3 \times 10^{14}$ |

**Notetaking with Vocabulary**

**Vocabulary:**

**Notes:**

## 14.5 Self-Assessment

Use the scale below to rate your understanding of the learning target and the success criteria.

| **1** | **2** | **3** | **4** |
|---|---|---|---|
| I do not understand. | I can do it with help. | I can do it on my own. | I can teach someone else. |

|  | Rating | Date |
|---|---|---|
| **14.5 Estimating Quantities** | | |
| **Learning Target:** Round numbers and write the results as the product of a single digit and a power of 10. | 1   2   3   4 | |
| I can round very large and very small numbers. | 1   2   3   4 | |
| I can write a multiple of 10 as a power. | 1   2   3   4 | |
| I can compare very large or very small quantities. | 1   2   3   4 | |

## 14.5 Practice

**Round the number. Write the result as a product of a single digit and a power of 10.**

1. 63,208,510,000

2. 45,007,899

3. Your neighbor is one of the winners of the lottery. Your neighbor's share of the winnings is $674,385. Write the result as a product of a single digit and a power of 10.

**Round the number. Write the result as a product of a single digit and a power of 10.**

4. 0.000000528

5. 0.00000000007398

6. The files on your computer contain about 3,894,467 words. The files on your friend's computer contain about 2.5 times the number of words as the files on your computer. What is the approximate number of words on your friend's computer?

7. The distance of a marathon running race is about 42,164.81 meters. The distance of a Ragnar relay race is about 7.5 times this distance.

   a. What is the approximate distance of a Ragnar relay race?

   b. A Ragnar relay team consists of 6 runners. If each runner runs the same distance, approximately how far will each runner run?

   c. Write the result in part (b) as the product of a single digit and a power of 10.

8. Is $7 \times 10^{-5}$ a better approximation of 0.00006973 or 0.00007271? Explain.

9. The surface area of Lake Ontario is about 204,070,000,000 square feet and the surface area of Lake Superior is about 883,745,000,000 square feet. Approximately how many times greater is the surface area of Lake Superior than the surface area of Lake Ontario?

10. Find a number that is 3.5 times 41,582,915,200. Write the result as the product of a single digit and a power of 10.

11. Your phone has about 784,392 bytes of storage left. Is this more or less than 1 megabyte of storage? (1 megabyte = $10^6$) Explain.

## 14.6 Scientific Notation
**For use with Exploration 14.6**

**Learning Target:** Understand the concept of scientific notation.

**Success Criteria:**
- I can convert between scientific notation and standard form.
- I can choose appropriate units to represent quantities.
- I can use scientific notation to solve real-life problems.

### 1 EXPLORATION: Using a Graphing Calculator

**Work with a partner. Use a graphing calculator.**

a. Experiment with multiplying very large numbers until your calculator displays an answer that is *not* in standard form. What do you think the answer means?

b. Enter the equation $y = 10^x$ into your graphing calculator. Use the *table* feature to find $y$-values for positive integer values of $x$ until the calculator displays a $y$-value that is not in standard form. Do the results support your answer in part (a)? Explain.

**14.6** **Scientific Notation** (continued)

   **c.** Repeat part (a) with very small numbers.

   **d.** Enter the equation $y = \left(\frac{1}{10}\right)^x$ into your graphing calculator. Use the *table* feature to find $y$-values for positive integer values of $x$ until the calculator displays a $y$-value that is not in standard form. Do the results support your answer in part (c)? Explain.

## 14.6 Notetaking with Vocabulary

**Vocabulary:**

**Notes:**

## 14.6 Self-Assessment

Use the scale below to rate your understanding of the learning target and the success criteria.

| *1* | *2* | *3* | *4* |
|---|---|---|---|
| I do not understand. | I can do it with help. | I can do it on my own. | I can teach someone else. |

| | Rating | Date |
|---|---|---|
| **14.6 Scientific Notation** | | |
| **Learning Target:** Understand the concept of scientific notation. | 1  2  3  4 | |
| I can convert between scientific notation and standard form. | 1  2  3  4 | |
| I can choose appropriate units to represent quantities. | 1  2  3  4 | |
| I can use scientific notation to solve real-life problems. | 1  2  3  4 | |

## 14.6 Practice

**Write the number in scientific notation.**

1. 0.000085

2. 410,000,000

3. 7,000,000,000,000,000

4. 0.00000000000199

**Write the number in standard form.**

5. $5 \times 10^{-4}$

6. $1.54 \times 10^5$

7. $1.78 \times 10^{-6}$

8. $3.555 \times 10^8$

9. The radius of Earth is about $6.38 \times 10^6$ meters. The radius of the Moon is about $1.74 \times 10^6$ meters. The radius of the Sun is about $7 \times 10^8$ meters.

   a. Which is the largest, *Earth*, the *Moon*, or the *Sun*?

   b. Which is the smallest, *Earth*, the *Moon*, or the *Sun*?

   c. Write the radius of Earth in standard form.

   d. Write the radius of the Moon in standard form.

   e. Write the radius of the Sun in standard form.

10. A year is about $3.156 \times 10^7$ seconds.

    a. How many seconds are in 5 years? Write your answer in standard form.

    b. How many seconds are in 1 month? Write your answer in standard form.

11. Approximately how many moons would be needed side-by-side to span across the Sun?

    Sun

    Moon

    $1.392 \times 10^6$ km    $3.475 \times 10^3$ km

    *not drawn to scale*

12. Most golf balls have about 250 to 450 dimples. The record holder is a ball with 1070 dimples.

    a. Write 1070 in scientific notation.

    b. In a recent year, it was estimated that 540,000,000 golf balls were sold. Using an average of 350 dimples, how many dimples were on the golf balls sold in that year? Write your answer in scientific notation.

## 14.7 Operations in Scientific Notation
**For use with Exploration 14.7**

**Learning Target:** Perform operations with numbers written in scientific notation.

**Success Criteria:**
- I can explain how to add and subtract numbers in scientific notation.
- I can explain how to multiply and divide numbers in scientific notation.
- I can use operations in scientific notation to solve real-life problems.

### 1 EXPLORATION: Adding and Subtracting in Scientific Notation

**Work with a partner.**

a. Complete the table by finding the sum and the difference of Expression 1 and Expression 2. Write your answers in scientific notation. Explain your method.

| Expression 1 | Expression 2 | Sum | Difference |
|---|---|---|---|
| $3 \times 10^4$ | $1 \times 10^4$ | | |
| $4 \times 10^{-3}$ | $2 \times 10^{-3}$ | | |
| $4.1 \times 10^{-7}$ | $1.5 \times 10^{-7}$ | | |
| $8.3 \times 10^6$ | $1.5 \times 10^6$ | | |

b. Use your results in part (a) to explain how to find $(a \times 10^n) + (b \times 10^n)$ and $(a \times 10^n) - (b \times 10^n)$.

**14.7** Operations in Scientific Notation (continued)

2 **EXPLORATION:** Multiplying and Dividing in Scientific Notation

**Work with a partner.**

a. Complete the table by finding the product and the quotient of Expression 1 and Expression 2. Write your answers in scientific notation. Explain your method.

| Expression 1 | Expression 2 | Product | Quotient |
|:---:|:---:|:---:|:---:|
| $3 \times 10^4$ | $1 \times 10^4$ | | |
| $4 \times 10^3$ | $2 \times 10^2$ | | |
| $7.7 \times 10^{-2}$ | $1.1 \times 10^{-3}$ | | |
| $4.5 \times 10^5$ | $3 \times 10^{-1}$ | | |

b. Use your results in part (a) to explain how to find $(a \times 10^n) \times (b \times 10^m)$ and $(a \times 10^n) \div (b \times 10^m)$. Describe any properties that you use.

## Notetaking with Vocabulary

**Vocabulary:**

**Notes:**

## 14.7 Self-Assessment

**Use the scale below to rate your understanding of the learning target and the success criteria.**

| 1 | 2 | 3 | 4 |
|---|---|---|---|
| I do not understand. | I can do it with help. | I can do it on my own. | I can teach someone else. |

|  | Rating | Date |
|---|:---:|---|
| **14.7 Operations in Scientific Notation** | | |
| **Learning Target:** Perform operations with numbers written in scientific notation. | 1   2   3   4 | |
| I can explain how to add and subtract numbers in scientific notation. | 1   2   3   4 | |
| I can explain how to multiply and divide numbers in scientific notation. | 1   2   3   4 | |
| I can use operations in scientific notation to solve real-life problems. | 1   2   3   4 | |

**Big Ideas Math: Modeling Real Life Grade 7 Accelerated**
Student Journal
**361**

Name _____ Date _____

## 14.7 Practice

**Find the sum or difference. Write your answer in scientific notation.**

1. $(1.4 \times 10^2) - (1.1 \times 10^2)$

2. $(5.2 \times 10^{-4}) - (4.58 \times 10^{-4})$

3. $(6.4 \times 10^{-2}) + (4.7 \times 10^{-3})$

4. $(5.92 \times 10^{14}) - (3 \times 10^{12})$

**Find the product or quotient. Write your answer in scientific notation.**

5. $(7.5 \times 10^{-5}) \div (3 \times 10^{-3})$

6. $(6.1 \times 10^{-6}) \times (3 \times 10^{-1})$

7. $(6.8 \times 10^{-14}) \div (8.5 \times 10^{10})$

8. $(6 \times 10^{-8}) \times (3.1 \times 10^{12})$

**Find the area of the figure. Write your answer in scientific notation.**

9.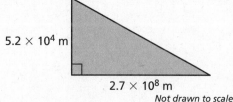

5.2 × 10⁴ m

2.7 × 10⁸ m
Not drawn to scale

10.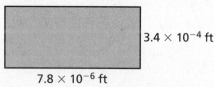

3.4 × 10⁻⁴ ft

7.8 × 10⁻⁶ ft
Not drawn to scale

11. How many times greater is the total area of Russia than the total area of Finland?

Finland
Total Area ≈ 3.4 × 10⁵ km²

Russia
Total Area ≈ 1.7 × 10⁷ km²

**Evaluate the expression. Write your answer in scientific notation.**

12. $48{,}000{,}000 \div (1.6 \times 10^3) + (2.7 \times 10^4)$

13. $(1.9 \times 10^9) - 6{,}300{,}000 \times (5.6 \times 10^2)$

14. You use technology and find a sum of 6.712E–8. Write this sum in standard form.

# Chapter 14 Chapter Self-Assessment

**Use the scale below to rate your understanding of the learning target and the success criteria.**

| 1 | 2 | 3 | 4 |
|---|---|---|---|
| I do not understand. | I can do it with help. | I can do it on my own. | I can teach someone else. |

| | Rating | Date |
|---|---|---|
| **14.1 Exponents** | | |
| **Learning Target:** Use exponents to write and evaluate expressions. | 1　2　3　4 | |
| I can write products using exponents. | 1　2　3　4 | |
| I can evaluate expressions involving powers. | 1　2　3　4 | |
| I can use exponents to solve real-life problems. | 1　2　3　4 | |
| **14.2 Product of Powers Property** | | |
| **Learning Target:** Generate equivalent expressions involving products of powers. | 1　2　3　4 | |
| I can find products of powers that have the same base. | 1　2　3　4 | |
| I can find powers of powers. | 1　2　3　4 | |
| I can find powers of products. | 1　2　3　4 | |
| **14.3 Quotient of Powers Property** | | |
| **Learning Target:** Generate equivalent expressions involving quotients of powers. | 1　2　3　4 | |
| I can find quotients of powers that have the same base. | 1　2　3　4 | |
| I can simplify expressions using the Quotient of Powers Property. | 1　2　3　4 | |
| I can solve real-life problems involving quotients of powers. | 1　2　3　4 | |

## Chapter 14 Chapter Self-Assessment (continued)

| | Rating | Date |
|---|---|---|
| **14.4 Zero and Negative Exponents** | | |
| **Learning Target:** Understand the concepts of zero and negative exponents. | 1  2  3  4 | |
| I can explain the meanings of zero and negative exponents. | 1  2  3  4 | |
| I can evaluate numerical expressions involving zero and negative exponents. | 1  2  3  4 | |
| I can simplify algebraic expressions involving zero and negative exponents. | 1  2  3  4 | |
| **14.5 Estimating Quantities** | | |
| **Learning Target:** Round numbers and write the results as the product of a single digit and a power of 10. | 1  2  3  4 | |
| I can round very large and very small numbers. | 1  2  3  4 | |
| I can write a multiple of 10 as a power. | 1  2  3  4 | |
| I can compare very large or very small quantities. | 1  2  3  4 | |
| **14.6 Scientific Notation** | | |
| **Learning Target:** Understand the concept of scientific notation. | 1  2  3  4 | |
| I can convert between scientific notation and standard form. | 1  2  3  4 | |
| I can choose appropriate units to represent quantities. | 1  2  3  4 | |
| I can use scientific notation to solve real-life problems. | 1  2  3  4 | |
| **14.7 Operations in Scientific Notation** | | |
| **Learning Target:** Perform operations with numbers written in scientific notation. | 1  2  3  4 | |
| I can explain how to add and subtract numbers in scientific notation. | 1  2  3  4 | |
| I can explain how to multiply and divide numbers in scientific notation. | 1  2  3  4 | |
| I can use operations in scientific notation to solve real-life problems. | 1  2  3  4 | |

**Chapter 15** **Review & Refresh**

**Complete the number sentence with <, >, or =.**

1. 3.4 _____ 3.45

2. −6.01 _____ −6.1

3. 3.50 _____ 3.5

4. −0.84 _____ −0.91

**Find three decimals that make the number sentence true.**

5. −5.2 ≥ _____

6. 2.65 > _____

7. −3.18 ≤ _____

8. 0.03 < _____

9. The table shows the times of a 100-meter dash. Order the runners from first place to fifth place.

| Runner | Time (seconds) |
|--------|----------------|
| A      | 12.60          |
| B      | 12.55          |
| C      | 12.49          |
| D      | 12.63          |
| E      | 12.495         |

# Chapter 15 Review & Refresh (continued)

**Evaluate the expression.**

**10.** $10^2 - 48 \div 6 + 25 \cdot 3$

**11.** $8\left(\frac{16}{4}\right) + 2^2 - 11 \cdot 3$

**12.** $\left(\frac{6}{3} + 4\right)^2 \div 4 \cdot 7$

**13.** $5(9 - 4)^2 - 3^2$

**14.** $5^2 - 2^2 \cdot 4^2 - 12$

**15.** $\left(\frac{50}{5^2}\right)^2 \div 4$

**16.** The table shows the numbers of students in 4 classes. The teachers are combining the classes and dividing the students in half to form two groups for a project. Write an expression to represent this situation. How many students are in each group?

| Class | Students |
|-------|----------|
| 1 | 24 |
| 2 | 32 |
| 3 | 30 |
| 4 | 28 |

## 15.1 Finding Square Roots
**For use with Exploration 15.1**

**Learning Target:** Understand the concept of a square root of a number.

**Success Criteria:**
- I can find square roots of numbers.
- I can evaluate expressions involving square roots.
- I can use square roots to solve equations.

**1 EXPLORATION: Finding Side Lengths**

**Work with a partner. Find the side length $s$ of each square. Explain your method.**

Area = 81 yd$^2$

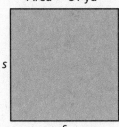

Area = 324 cm$^2$

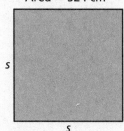

Area = 361 mi$^2$

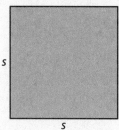

Area = 225 mi$^2$

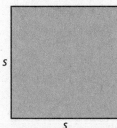

Area = 2.89 in.$^2$

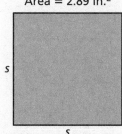

Area = $\frac{4}{9}$ ft$^2$

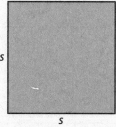

**15.1** **Finding Square Roots** (continued)

---

**2** **EXPLORATION:** Finding Solutions of Equations

**Work with a partner. Use mental math to solve each equation. How many solutions are there for each equation? Explain your reasoning.**

$$x^2 = 0$$

$$x^2 = 1$$

$$x^2 = 4$$

$$x^2 = 9$$

$$x^2 = 16$$

$$x^2 = 25$$

---

## 15.1 Notetaking with Vocabulary

**Vocabulary:**

**Notes:**

## 15.1 Self-Assessment

**Use the scale below to rate your understanding of the learning target and the success criteria.**

| **1** | **2** | **3** | **4** |
|---|---|---|---|
| I do not understand. | I can do it with help. | I can do it on my own. | I can teach someone else. |

|  | Rating | Date |
|---|---|---|
| **15.1 Finding Square Roots** | | |
| **Learning Target:** Understand the concept of a square root of a number. | 1  2  3  4 | |
| I can find square roots of numbers. | 1  2  3  4 | |
| I can evaluate expressions involving square roots. | 1  2  3  4 | |
| I can use square roots to solve equations. | 1  2  3  4 | |

Name _____ Date _____

## 15.1 Practice

**Find the dimensions of the square or circle.**

1. Area = $\dfrac{169}{225}$ cm$^2$

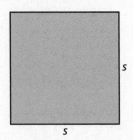

2. Area = $121\pi$ yd$^2$

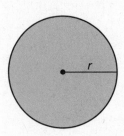

3. You have a square plot of land that has an area of 144 square feet. You want to plant the largest circular flower garden possible on the plot of land. What is the area of the flower garden? Use 3.14 for $\pi$.

**Find the square root(s).**

4. $-\sqrt{484}$

5. $\pm\sqrt{\dfrac{25}{64}}$

6. A cylindrical oil tank holds about 785 cubic feet of oil and has a height of 10 feet. The formula for the volume of a cylinder is $V = \pi r^2 h$. Find the radius of the tank. Use 3.14 for $\pi$.

**Evaluate the expression.**

7. $6\sqrt{2.25} - 4.2$

8. $3\left(\sqrt{\dfrac{48}{3}} - 2\right)$

**Copy and complete the statement with <, >, or =.**

9. $\sqrt{\dfrac{49}{9}}$ _____ 2

10. $\dfrac{2}{5}$ _____ $\sqrt{\dfrac{12}{75}}$

11. The area of a sector of a circle is represented by $A = \dfrac{5}{18}\pi r^2$, where $r$ is the radius of the circle (in meters). What is the radius when the area is $40\pi$ square meters?

12. Two squares are drawn. The smaller square has an area of 256 square meters. The areas of the two squares have a ratio of 4 : 9. What is the side length $s$ of the larger square?

13. The cost $C$ (in dollars) of producing $x$ DVD players is represented by $C = 4.5x^2$. How many DVD players are produced if the cost is $544.50?

## 15.2 The Pythagorean Theorem

**For use with Exploration 15.2**

**Learning Target:**   Understand the Pythagorean Theorem.

**Success Criteria:**
- I can explain the Pythagorean Theorem.
- I can use the Pythagorean Theorem to find unknown side lengths of triangles.
- I can use the Pythagorean Theorem to find distances between points in a coordinate plane.

---

**1   EXPLORATION: Discovering the Pythagorean Theorem**

**Work with a partner.**

- On grid paper, draw a right triangle with one horizontal side and one vertical side.

- Label the lengths of the two shorter sides *a* and *b*. Label the length of the longest side *c*.

- Draw three squares that each share a side with your triangle. Label the areas of the squares $a^2, b^2$, and $c^2$.

- Cut out each square. Then make eight copies of the right triangle and cut them out (*).

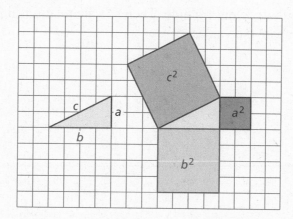

* Figures are available in the back of the Student Journal.

---

**15.2** **The Pythagorean Theorem** (continued)

**a.** Arrange the figures to show how $a^2$ and $b^2$ relate to $c^2$. Use an equation to represent this relationship.

**b.** Estimate the side length $c$ of your triangle. Then use the relationship in part (a) to find $c$. Compare the values.

# 15.2 Notetaking with Vocabulary

**Vocabulary:**

**Notes:**

# 15.2 Self-Assessment

Use the scale below to rate your understanding of the learning target and the success criteria.

| **1** | **2** | **3** | **4** |
|---|---|---|---|
| I do not understand. | I can do it with help. | I can do it on my own. | I can teach someone else. |

| | Rating | Date |
|---|:---:|:---:|
| **15.2 The Pythagorean Theorem** | | |
| **Learning Target:** Understand the Pythagorean Theorem. | 1   2   3   4 | |
| I can explain the Pythagorean Theorem. | 1   2   3   4 | |
| I can use the Pythagorean Theorem to find unknown side lengths of triangles. | 1   2   3   4 | |
| I can use the Pythagorean Theorem to find distances between points in a coordinate plane. | 1   2   3   4 | |

## 15.2 Practice

**Find the missing length of the triangle.**

1.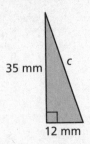

   35 mm

   c

   12 mm

2.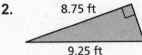

   8.75 ft

   a

   9.25 ft

3. You built braces in the shape of a right triangle to hold your surfboard. The leg (brace) attached to the wall is 10 inches and your surfboard sits on a leg that is 24 inches. What is the length of the hypotenuse that completes the right triangle?

4. Laptops are advertised by the lengths of the diagonals of the screen. You purchase a 15-inch laptop and the width of the screen is 12 inches. What is the height of its screen?

5. In a right isosceles triangle, the lengths of both legs are equal. For the given isosceles triangle, what is the value of $x$?

   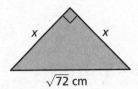

   $x$   $x$

   $\sqrt{72}$ cm

6. To get from your house to your school, you ride your bicycle 6 blocks west and 8 blocks north. A new road is being built that will go directly from your house to your school, creating a right triangle. When you take the new road to school, how many fewer blocks will you be riding to school and back?

   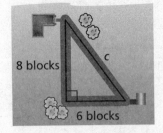

   8 blocks

   c

   6 blocks

7. The points A, B and C are the vertices of a triangle. The coordinates of the points are A(5,3), B(5,6) and C($x$,3). Find two values for $x$ so the points form a right triangle.

8. The legs of a right triangle have lengths of 15 feet and 36 feet. The hypotenuse has a length of $13x$ feet. What is the value of $x$?

9. You have a box that is a cube with side length of 10 cm. Will a pencil that is 17 cm long fit inside the box? Explain your reasoning.

Name_____ Date_____

## 15.3 Finding Cube Roots
**For use with Exploration 15.3**

**Learning Target:** Understand the concept of a cube root of a number.

**Success Criteria:**
- I can find cube roots of numbers.
- I can evaluate expressions involving cube roots.
- I can use cube roots to solve equations.

---

**1 EXPLORATION:** Finding Edge Lengths

**Work with a partner. Find the edge length $s$ of each cube. Explain your method.**

Volume = 8 cm³

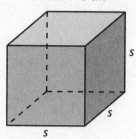

Volume = 27 ft³

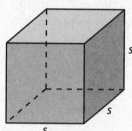

Volume = 125 m³

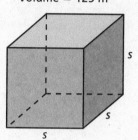

Volume = 343 in.³

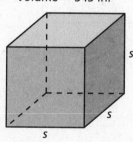

Volume = 0.001 cm³

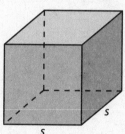

Volume = $\frac{1}{8}$ yd³

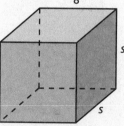

**15.3** **Finding Cube Roots** (continued)

---

**2** **EXPLORATION:** Finding Solutions of Equations

**Work with a partner. Use mental math to solve each equation. How many solutions are there for each equation? Explain your reasoning.**

$$x^3 = -27$$

$$x^3 = -8$$

$$x^3 = -1$$

$$x^3 = 1$$

$$x^3 = 8$$

$$x^3 = 27$$

---

## 15.3 Notetaking with Vocabulary

**Vocabulary:**

**Notes:**

## 15.3 Self-Assessment

Use the scale below to rate your understanding of the learning target and the success criteria.

| 1 | 2 | 3 | 4 |
|---|---|---|---|
| I do not understand. | I can do it with help. | I can do it on my own. | I can teach someone else. |

| | Rating | Date |
|---|---|---|
| **15.3 Finding Cube Roots** | | |
| **Learning Target:** Understand the concept of a cube root of a number. | 1  2  3  4 | |
| I can find cube roots of numbers. | 1  2  3  4 | |
| I can evaluate expressions involving cube roots. | 1  2  3  4 | |
| I can use cube roots to solve equations. | 1  2  3  4 | |

Name _____ Date _____

## 15.3 Practice

**Find the cube root.**

1. $\sqrt[3]{343}$

2. $\sqrt[3]{-1331}$

3. $\sqrt[3]{-\dfrac{125}{27}}$

**Evaluate the expression.**

4. $13 + \left(\sqrt[3]{125}\right)^3$

5. $2\dfrac{2}{3} - \left(\sqrt[3]{\dfrac{1}{27}}\right)^3$

6. $24 + \left(\sqrt[3]{-1000}\right)^3$

**Evaluate the expression for the given value of the variable.**

7. $\sqrt[3]{4t} + 3t, t = 54$

8. $\sqrt[3]{\dfrac{n}{24}} - \dfrac{n}{25}, n = 375$

9. The volume of a storage pod that is shaped like a cube is 1728 cubic feet.

   a. What is the edge length of the storage pod?

   b. What is the surface area of the storage pod?

   c. What is the area of the floor space of the storage pod?

**Copy and complete the statement with <, >, or =.**

10. $0.25 \_\_\_\_ \sqrt[3]{0.008}$

11. $\sqrt{729} \_\_\_\_ \sqrt[3]{729}$

12. There are infinitely many pairs of numbers of which the sum of their cube roots is zero. Give two of these pairs.

13. The radius of a sphere can be represented by $r = \sqrt[3]{\dfrac{3V}{4\pi}}$, where $V$ is the volume of the sphere. What is the radius of a sphere with a volume of $36\pi$ cubic meters?

**Solve the equation.**

14. $(4x - 1)^3 = 343$

15. $(15x^3 - 2)^3 = 2197$

16. The ratio $x : 8$ is equivalent to the ratio $27 : x^2$. What is the value of $x$?

## 15.4 Rational Numbers
### For use with Exploration 15.4

**Learning Target:** Convert between different forms of rational numbers.

**Success Criteria:** 
- I can explain the meaning of rational numbers.
- I can write fractions and mixed numbers as decimals.
- I can write repeating decimals as fractions or mixed numbers.

---

**1 EXPLORATION: Writing Repeating Decimals as Fractions**

**Work with a partner.**

**a.** Complete the table.

| $x$ | $10x$ |
|---|---|
| $x = 0.333\ldots$ | $10x = 3.333\ldots$ |
| $x = 0.666\ldots$ | |
| $x = 0.111\ldots$ | |
| $x = 0.2444\ldots$ | |

**b.** For each row of the table, use the two equations to write an equivalent equation that does not involve a repeating decimal. Then solve the equation. What does your solution represent?

**15.4** **Rational Numbers** (continued)

c. Write each repeating decimal below as a fraction. How is your procedure
similar to parts (a) and (b)? How is it different?

$x = 0.\overline{12}$

$x = 0.\overline{45}$

$x = 0.\overline{27}$

$x = 0.9\overline{40}$

d. Explain how to write a repeating decimal with $n$ repeating digits as a
fraction.

## 15.4 Notetaking with Vocabulary

**Vocabulary:**

**Notes:**

## 15.4 Self-Assessment

Use the scale below to rate your understanding of the learning target and the success criteria.

| **1** | **2** | **3** | **4** |
|---|---|---|---|
| I do not understand. | I can do it with help. | I can do it on my own. | I can teach someone else. |

| | Rating | Date |
|---|:---:|:---:|
| **15.4 Rational Numbers** | | |
| **Learning Target:** Convert between different forms of rational numbers. | 1   2   3   4 | |
| I can explain the meaning of rational numbers. | 1   2   3   4 | |
| I can write fractions and mixed numbers as decimals. | 1   2   3   4 | |
| I can write repeating decimals as fractions or mixed numbers. | 1   2   3   4 | |

Name _____ Date _____

## 15.4 Practice

**Write the fraction or mixed number as a decimal.**

1. $\dfrac{7}{36}$

2. $-7\dfrac{4}{75}$

3. $3\dfrac{2}{11}$

4. The length of your computer mouse is $4\dfrac{7}{16}$ inches long. Write this length as a decimal.

**Write the repeating decimal as a fraction or a mixed number.**

5. $8.\overline{7}$

6. $24.\overline{8}$

7. $-1.4\overline{5}$

8. $-0.\overline{32}$

9. $6.\overline{13}$

10. $7.\overline{90}$

11. The probability of rolling a 5 when rolling a 6-sided die is $0.1\overline{66}$. Write this probability as a fraction.

12. You are making two types of cookies for the math club bake sale.

   a. The recipe for Cookie A uses 0.6 times the amount of flour used in the recipe for Cookie B. The recipe for Cookie B calls for $3\dfrac{2}{3}$ cups of flour. How many cups of flour should you buy to have enough for both recipes?

   b. Both recipes call for $0.\overline{66}$ cups of brown sugar. How many cups of brown sugar should you buy to have enough for both recipes? Write this number as a fraction.

   c. The recipe for Cookie B uses $\dfrac{3}{4}$ times the amount of butter used in the recipe for Cookie A. The recipe for Cookie A calls for $1.08\overline{33}$ cups of butter. How many cups of butter should you buy to have enough for both recipes? Write this number as a decimal.

**Add or subtract.**

13. $0.3\overline{05} + 0.2\overline{41}$

14. $\dfrac{7}{44} - 0.\overline{99}$

15. $0.\overline{01} - 0.\overline{05}$

16. Write a repeating decimal that is between $\dfrac{3}{11}$ and $\dfrac{4}{11}$. Justify your answer.

17. Write two repeating decimals that $\dfrac{17}{30}$ falls between. Justify your answer.

**Determine whether the numbers are equal. Justify your answer.**

18. $\dfrac{7}{33}$ and 0.21

19. $\dfrac{5}{44}$ and $0.11\overline{36}$

20. $\dfrac{6}{55}$ and $0.1\overline{09}$

## 15.5  Irrational Numbers
**For use with Exploration 15.5**

**Learning Target:**   Understand the concept of irrational numbers.

**Success Criteria:**   • I can classify real numbers as rational or irrational.
  • I can approximate irrational numbers.
  • I can solve real-life problems involving irrational numbers.

---

**1  EXPLORATION: Approximating Square Roots**

**Work with a partner. Use the square shown.**

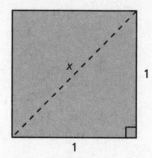

**a.** Find the exact length $x$ of the diagonal. Is this number a *rational number* or an *irrational number*? Explain.

**b.** The value of $x$ is between which two whole numbers? Explain your reasoning.

**Big Ideas Math: Modeling Real Life Grade 7 Accelerated**   **383**

**15.5** **Irrational Numbers** (continued)

**c.** Use the diagram below (*) to approximate the length of the diagonal to the nearest tenth. Explain your method.

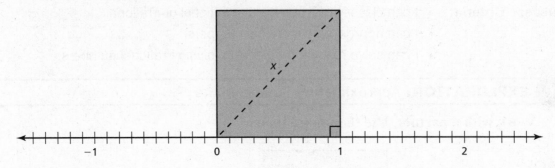

**d.** Which of the following is the closest approximation of the length of the diagonal? Justify your answer using inverse operations.

| 1.412 | 1.413 | 1.414 | 1.415 |

* Diagram is available in the back of the Student Journal.

## 15.5 Notetaking with Vocabulary

**Vocabulary:**

**Notes:**

## 15.5 Self-Assessment

Use the scale below to rate your understanding of the learning target and the success criteria.

| **1** | **2** | **3** | **4** |
|---|---|---|---|
| I do not understand. | I can do it with help. | I can do it on my own. | I can teach someone else. |

| | Rating | Date |
|---|---|---|
| **15.5 Irrational Numbers** | | |
| **Learning Target:** Understand the concept of irrational numbers. | 1  2  3  4 | |
| I can classify real numbers as rational or irrational. | 1  2  3  4 | |
| I can approximate irrational numbers. | 1  2  3  4 | |
| I can solve real-life problems involving irrational numbers. | 1  2  3  4 | |

## 15.5 Practice

**Classify the real number.**

1. $2\frac{2}{9}$

2. $-\sqrt{576}$

3. $2.\overline{41}$

4. $\sqrt{130}$

5. You are finding the circumference of a circle with a diameter of 10 meters. Is the circumference a *rational* or *irrational* number? Explain.

**Approximate the number to the nearest (a) integer and (b) tenth.**

6. $-\sqrt{\dfrac{250}{9}}$

7. $\sqrt{395}$

8. $\sqrt[3]{-50}$

9. $\sqrt{0.79}$

10. $\sqrt{1.48}$

11. $\sqrt[3]{294}$

12. A patio is in the shape of a square, with a side length of 35 feet. You wish to draw a black line down one diagonal.

   a. Use the Pythagorean Theorem to find the length of the diagonal. Write your answer as a square root.

   b. Find the two perfect squares that the length of the diagonal falls between.

   c. Estimate the length of the diagonal to the nearest tenth.

**Which number is greater? Explain.**

13. $\sqrt{220}, \sqrt[3]{1260}$

14. $-\sqrt{135}, -\sqrt{145}$

15. $\sqrt{\dfrac{7}{64}}, \sqrt[3]{\dfrac{1}{28}}$

16. $2\pi, \sqrt[3]{250}$

17. Find two numbers $a$ and $b$ such that $7 < \sqrt{a} < \sqrt{b} < 8$.

18. Is $\sqrt{0.0625}$ a rational number? Explain.

19. The volume of a cube is 88 cubic centimeters.

   a. Approximate the side length $s$ of the cube to the nearest whole number.

   b. Approximate the side length $s$ of the cube to the nearest tenth.

## 15.6 The Converse of the Pythagorean Theorem
**For use with Exploration 15.6**

**Learning Target:** Understand the converse of the Pythagorean Theorem.

**Success Criteria:**
- I can explain the converse of the Pythagorean Theorem.
- I can identify right triangles given three side lengths.
- I can identify right triangles in a coordinate plane.

---

**1  EXPLORATION:** Analyzing the Converse of a Statement

**Work with a partner.**

**a.** Write the converse of each statement. Then determine whether each statement and its converse are *true* or *false*. Explain.

- If I live in California, then I live in the United States.

- If my heart is beating, then I am alive.

- If one figure is a translation of another figure, then the figures are congruent.

**b.** Write your own statement whose converse is true. Then write your own statement whose converse is false.

**15.6** **The Converse of the Pythagorean Theorem** (continued)

---

**2** **EXPLORATION:** The Converse of the Pythagorean Theorem

**Work with a partner.**

**a.** Write the converse of the Pythagorean Theorem. Do you think the converse is *true* or *false*?

**b.** Consider $\triangle DEF$ with side lengths $a$, $b$, and $c$ such that $a^2 + b^2 = c^2$. Also consider $\triangle JKL$ with leg lengths $a$ and $b$, where the measure of $\angle K$ is $90°$. Use the two triangles and the Pythagorean Theorem to show that the converse of the Pythagorean Theorem is true.

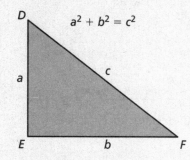
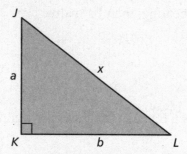

## 15.6 Notetaking with Vocabulary

**Vocabulary:**

**Notes:**

## 15.6 Self-Assessment

Use the scale below to rate your understanding of the learning target and the success criteria.

| *1* | *2* | *3* | *4* |
|---|---|---|---|
| I do not understand. | I can do it with help. | I can do it on my own. | I can teach someone else. |

| | Rating | Date |
|---|---|---|
| **15.6 The Converse of the Pythagorean Theorem** | | |
| **Learning Target:** Understand the converse of the Pythagorean Theorem. | 1   2   3   4 | |
| I can explain the converse of the Pythagorean Theorem. | 1   2   3   4 | |
| I can identify right triangles given three side lengths. | 1   2   3   4 | |
| I can identify right triangles in a coordinate plane. | 1   2   3   4 | |

Name _____ Date _____

# 15.6 Practice

**Write the converse of the statement. Then determine whether the statement and its converse are *true* or *false*. Explain.**

1. If you live in New York City, then you live in New York.

2. If $a$ is a perfect square, then $\sqrt{a}$ is an integer.

**Tell whether the triangle with the given side lengths is a right triangle.**

3. 11 in., 60 in., 61 in.

4. 45 cm, 26 cm, 51 cm

5. You are building an entrance sign to a resort. The entrance sign will have side lengths of 8.1 feet, 8.1 feet, and 11.5 feet. Is the sign a right triangle? Explain.

**Tell whether a triangle with the given side lengths is a right triangle.**

6. $9, \sqrt{54}, 8$

7. $\sqrt{704}, 27, 5$

8. $88, 103, 137$

9. You are creating a flower garden in the triangular shape shown. You purchase edging to go around the flower garden. The edging costs $1.50 per foot. What is the cost of the edging? Round your lengths to the nearest whole number.

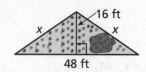

10. You and your friend are volunteering at the local preschool tricycle race. The locations of the markers for the race are represented by the points $A(0, 0)$, $B(14, 4)$, and $C(20, -17)$. Each unit represents 1 foot.

   a. Do the three markers, $A$, $B$, and $C$, form a right triangle?

   b. The fastest preschooler pedals 6 feet per minute. How fast does the fastest preschooler take to pedal around the entire triangle?

   c. A preschooler accidentally moves the $B$ marker to the $(12, 12)$ location. Do the three markers still form a right triangle? Explain.

   d. You and your friend decide to shorten the original course for a running race. The course is shortened by dividing the $x$- and $y$-coordinates of the three markers by 2. What are the new points representing the markers?

   e. Do the three markers resulting from part (d) form a right triangle? Explain.

# Chapter Self-Assessment

Use the scale below to rate your understanding of the learning target and the success criteria.

| 1 | 2 | 3 | 4 |
|---|---|---|---|
| I do not understand. | I can do it with help. | I can do it on my own. | I can teach someone else. |

|  | Rating | Date |
|---|---|---|
| **15.1 Finding Square Roots** | | |
| **Learning Target:** Understand the concept of a square root of a number. | 1  2  3  4 | |
| I can find square roots of numbers. | 1  2  3  4 | |
| I can evaluate expressions involving square roots. | 1  2  3  4 | |
| I can use square roots to solve equations. | 1  2  3  4 | |
| **15.2 The Pythagorean Theorem** | | |
| **Learning Target:** Understand the Pythagorean Theorem. | 1  2  3  4 | |
| I can explain the Pythagorean Theorem. | 1  2  3  4 | |
| I can use the Pythagorean Theorem to find unknown side lengths of triangles. | 1  2  3  4 | |
| I can use the Pythagorean Theorem to find distances between points in a coordinate plane. | 1  2  3  4 | |
| **15.3 Finding Cube Roots** | | |
| **Learning Target:** Understand the concept of a cube root of a number. | 1  2  3  4 | |
| I can find cube roots of numbers. | 1  2  3  4 | |
| I can evaluate expressions involving cube roots. | 1  2  3  4 | |
| I can use cube roots to solve equations. | 1  2  3  4 | |

## Chapter 15 — Chapter Self-Assessment (continued)

| | Rating | Date |
|---|---|---|
| **15.4 Rational Numbers** | | |
| **Learning Target:** Convert between different forms of rational numbers. | 1   2   3   4 | |
| I can explain the meaning of rational numbers. | 1   2   3   4 | |
| I can write fractions and mixed numbers as decimals. | 1   2   3   4 | |
| I can write repeating decimals as fractions or mixed numbers. | 1   2   3   4 | |
| **15.5 Irrational Numbers** | | |
| **Learning Target:** Understand the concept of irrational numbers. | 1   2   3   4 | |
| I can classify real numbers as rational or irrational. | 1   2   3   4 | |
| I can approximate irrational numbers. | 1   2   3   4 | |
| I can solve real-life problems involving irrational numbers. | 1   2   3   4 | |
| **15.6 The Converse of the Pythagorean Theorem** | | |
| **Learning Target:** Understand the converse of the Pythagorean Theorem. | 1   2   3   4 | |
| I can explain the converse of the Pythagorean Theorem. | 1   2   3   4 | |
| I can identify right triangles given three side lengths. | 1   2   3   4 | |
| I can identify right triangles in a coordinate plane. | 1   2   3   4 | |

# Chapter 16 Review & Refresh

**Find the area of the figure.**

**1.**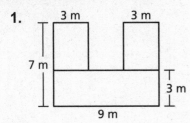

**2.**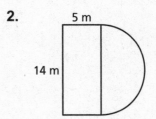

**3.**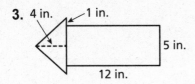

**4.**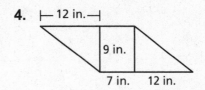

**5.**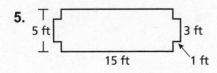

**6.**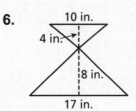

**7.** You are carpeting 2 rooms of your house. The carpet costs $1.48 per square foot. How much does it cost to carpet the rooms?

**Chapter 16** **Review & Refresh** (continued)

**Find the area of the circle.**

8.
20 in.

9.
6 m

10.
12 cm

11.
14 ft

12.
25 yd

13.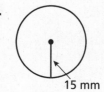
15 mm

14. Find the area of the shaded region.

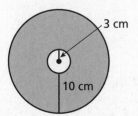
3 cm
10 cm

## 16.1 Volumes of Cylinders
### For use with Exploration 16.1

**Learning Target:** Find the volume of a cylinder.

**Success Criteria:**
- I can use a formula to find the volume of a cylinder.
- I can use the formula for the volume of a cylinder to find a missing dimension.

### 1 EXPLORATION: Exploring Volume

**Work with a partner.**

a. Each prism shown has a height of $h$ units and bases with areas of $B$ square units. Write a formula that you can use to find the volume of each prism.

Triangular Prism

Rectangular Prism

Pentagonal Prism

Hexagonal Prism

Octagonal Prism

b. How can you find the volume of a prism with bases that each have 100 sides?

c. Make a conjecture about how to find the volume of a cylinder. Explain your reasoning.

**16.1** **Volumes of Cylinders** (continued)

---

**2** **EXPLORATION:** Finding Volume Experimentally

**Work with a partner. Draw a net for a cylinder. Then cut out the net and use tape to form an open cylinder. Repeat this process to form an open cube. The edge length of the cube should be greater than the diameter and the height of the cylinder (\*).**

   **a.** Use your conjecture in Exploration 1 to find the volume of the cylinder.

   **b.** Fill the cylinder with rice. Then pour the rice into the open cube. Find the volume of rice in the cube. Does this support your answer in part (a)? Explain your reasoning.

   \* Nets are available in the back of the Student Journal.

## 16.1 Notetaking with Vocabulary

**Vocabulary:**

**Notes:**

## 16.1 Self-Assessment

Use the scale below to rate your understanding of the learning target and the success criteria.

| *1* | *2* | *3* | *4* |
|---|---|---|---|
| I do not understand. | I can do it with help. | I can do it on my own. | I can teach someone else. |

| | Rating | Date |
|---|---|---|
| **16.1 Volumes of Cylinders** | | |
| **Learning Target:** Find the volume of a cylinder. | 1  2  3  4 | |
| I can use a formula to find the volume of a cylinder. | 1  2  3  4 | |
| I can use the formula for the volume of a cylinder to find a missing dimension. | 1  2  3  4 | |

# 16.1 Practice

**Find the volume of the cylinder. Round your answer to the nearest tenth.**

**1.**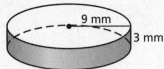

**2.**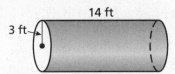

**3.** A cylinder has a surface are of 748 cm² and a radius of 7 cm. Estimate the volume of the cylinder to the nearest whole number.

**4.** A cylinder has a volume of 100π cubic meters.

   **a.** What is the volume of the cylinder if the height is halved? Explain.

   **b.** What is the volume of the cylinder if the diameter is halved? Explain.

**Find the missing dimension of the cylinder. Round your answer to the nearest whole number.**

**5.** Volume = 550 in.³

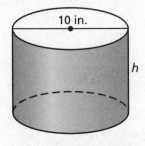

**6.** Volume = 25,000 ft³

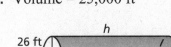
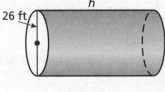

**7.** Your friend's swimming pool is in the shape of a rectangular prism, with a length of 25 feet, a width of 8 feet, and a height of 5 feet.

   **a.** What is the volume of your friend's swimming pool?

   **b.** Your swimming pool is in the shape of a cylinder with a diameter of 16 feet and has the same volume as your friend's pool. What is the height of your pool? Round your answer to the nearest whole number.

   **c.** While you were on vacation, 6 inches of water evaporated from your pool. About how many gallons of water evaporated from your pool? (1 ft³ ≈ 7.5 gal) Round your answer to the nearest whole number.

## 16.2 Volumes of Cones
**For use with Exploration 16.2**

**Learning Target:** Find the volume of a cone.

**Success Criteria:**
- I can use a formula to find the volume of a cone.
- I can use the formula for the volume of a cone to find a missing dimension.

---

**1 EXPLORATION:** Finding a Formula Experimentally

**Work with a partner. Use a paper cup that is shaped like a cone. Measure the height of the cup and the diameter of the circular base. Use these measurements to draw a net for a cylinder with the same base and height as the paper cup. Then cut out the net and use tape to form an open cylinder (\*).**

  **a.** Find the volume of the cylinder.

  **b.** Fill the paper cup with rice. Then pour the rice into the cylinder. Repeat this until the cylinder is full. How many cones does it take to fill the cylinder?

\* Nets are available in the back of the Student Journal.

**16.2** **Volumes of Cones** (continued)

**c.** Use your result to write a formula for the volume of a cone.

**d.** Use your formula in part (c) to find the volume of the cone. How can you tell whether your answer is correct?

**e.** Do you think your formula for the volume of a cone is also true for *oblique* cones? Explain your reasoning.

## 16.2 Notetaking with Vocabulary

**Vocabulary:**

**Notes:**

## 16.2 Self-Assessment

**Use the scale below to rate your understanding of the learning target and the success criteria.**

| 1 | 2 | 3 | 4 |
|---|---|---|---|
| I do not understand. | I can do it with help. | I can do it on my own. | I can teach someone else. |

| | Rating | Date |
|---|:---:|:---:|
| **16.2 Volumes of Cones** | | |
| **Learning Target:** Find the volume of a cone. | 1   2   3   4 | |
| I can use a formula to find the volume of a cone. | 1   2   3   4 | |
| I can use the formula for the volume of a cone to find a missing dimension. | 1   2   3   4 | |

# 16.2 Practice

**Find the volume of the cone. Round your answer to the nearest tenth.**

**1.**

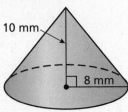

10 mm

8 mm

**2.**

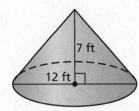

7 ft

12 ft

**3.** The volume of a cylinder is twice the volume of a cone. The cone and the cylinder have the same diameter. The height of the cylinder is 5 meters. What is the height of the cone?

**4.** One package of popcorn makes 1000 cubic inches of popcorn. The movie theatre sells the popcorn in cone shaped containers. The containers have a radius of 3 inches and a height of 8 inches. How many containers can the movie theatre fill from one package of popcorn?

**Find the missing dimension of the cone. Round your answer to the nearest tenth.**

**5.** Volume = 100 in.$^3$

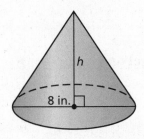

h

8 in.

**6.** Volume = 13.4 m$^3$

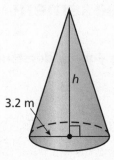

h

3.2 m

**7.** A paper cup is in the shape of a cone, with a diameter of 2 centimeters and a height of 5 centimeters.

   **a.** What is the volume of the paper cup?

   **b.** Water is running into the cup at a rate of 1.5 cubic centimeters per second. How long does it take for the cup to fill with water? Round your answer to the nearest tenth.

**8.** Cone A has the same radius but half the height of Cone B. What is the ratio of the volume of Cone A to the volume of Cone B?

## 16.3 Volumes of Spheres
**For use with Exploration 16.3**

**Learning Target:** Find the volume of a sphere.

**Success Criteria:**
- I can use a formula to find the volume of a sphere.
- I can use the formula for the volume of a sphere to find the radius.
- I can find volumes of composite solids.

---

**1 EXPLORATION: Finding a Formula Experimentally**

**Work with a partner. Use a plastic ball similar to the one shown. Draw a net for a cylinder with a diameter and a height equal to the diameter of the ball. Then cut out the net and use tape to form an open cylinder.**

**a.** How is the height $h$ of the cylinder related to the radius $r$ of the ball?

**16.3** **Volumes of Spheres** (continued)

**b.** Cover the ball with aluminum foil or tape. Leave one hole open. Fill the ball with rice. Then pour the rice into the cylinder. What fraction of the cylinder is filled with the rice?

**c.** Use your result in part (b) and the formula for the volume of a cylinder to write a formula for the volume of a sphere. Explain your reasoning.

## 16.3 Notetaking with Vocabulary

**Vocabulary:**

**Notes:**

## 16.3 Self-Assessment

**Use the scale below to rate your understanding of the learning target and the success criteria.**

| 1 | 2 | 3 | 4 |
|---|---|---|---|
| I do not understand. | I can do it with help. | I can do it on my own. | I can teach someone else. |

| | Rating | Date |
|---|---|---|
| **16.3 Volumes of Spheres** | | |
| **Learning Target:** Find the volume of a sphere. | 1  2  3  4 | |
| I can use a formula to find the volume of a sphere. | 1  2  3  4 | |
| I can use the formula for the volume of a sphere to find the radius. | 1  2  3  4 | |
| I can find volumes of composite solids. | 1  2  3  4 | |

## 16.3 Practice

**Find the volume of the sphere. Round your answer to the nearest tenth.**

1.
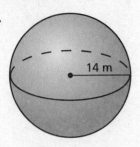
14 m

2.
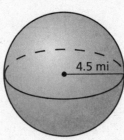
4.5 mi

3. A cylinder and a sphere have the same radius and volume. The height of the cylinder is 12 cm. What is the volume of the sphere rounded to the nearest whole number?

4. You want to give your friend a soccer ball for a gift. The soccer ball has a volume of $85.3\pi$ cubic inches. What is the volume of the smallest cube shaped box that the ball will fit inside?

**Find the radius of a sphere with the given volume. Round your answer to the nearest tenth if necessary.**

5. Volume = $2304\pi$ yd$^3$

6. Volume = $1543.5\pi$ yd$^3$

7. A spherical cabinet knob has a radius of 1.5 inches. Find the volume of the cabinet knob. Round your answer to the nearest tenth.

**Find the volume of the composite solid. Round your answer to the nearest tenth.**

8.
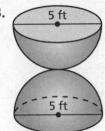
5 ft
5 ft

9.
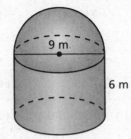
9 m
6 m

10. Rubber balls are packaged in a box with dimensions 8.5 cm, 5 cm, 5 cm. Each ball has a diameter of 3 cm. If there are a dozen balls in the box, how much space in the box is not occupied by the rubber balls? Round your answer to the nearest tenth.

## 16.4 Surface Areas and Volumes of Similar Solids
**For use with Exploration 16.4**

**Learning Target:** Find the surface areas and volumes of similar solids.

**Success Criteria:**
- I can use corresponding dimensions to determine whether solids are similar.
- I can use corresponding dimensions to find missing measures in similar solids.
- I can use linear measures to find surface areas and volumes of similar solids.

### 1 EXPLORATION: Comparing Similar Solids

**Work with a partner.**

**a.** You multiply the dimensions of the smallest cylinder by different factors to create the other four cylinders. Complete the table. Compare the surface area and volume of each cylinder with the surface area and volume of the smallest cylinder.

| Radius | 1 | 2 | 3 | 4 | 5 |
|---|---|---|---|---|---|
| Height | 1 | 2 | 3 | 4 | 5 |
| Surface Area | | | | | |
| Volume | | | | | |

**16.4** **Surface Areas and Volumes of Similar Solids** (continued)

**b.** Repeat part (a) using the square pyramids and table below.

| Base Side | 6 | 12 | 18 | 24 | 30 |
|---|---|---|---|---|---|
| Height | 4 | 8 | 12 | 16 | 20 |
| Slant Height | 5 | 10 | 15 | 20 | 25 |
| Surface Area | | | | | |
| Volume | | | | | |

## 16.4 Notetaking with Vocabulary

**Vocabulary:**

**Notes:**

## 16.4 Self-Assessment

**Use the scale below to rate your understanding of the learning target and the success criteria.**

| 1 | 2 | 3 | 4 |
|---|---|---|---|
| I do not understand. | I can do it with help. | I can do it on my own. | I can teach someone else. |

|  | Rating | Date |
|---|---|---|
| **16.4 Surface Areas and Volumes of Similar Solids** | | |
| **Learning Target:** Find the surface areas and volumes of similar solids. | 1  2  3  4 | |
| I can use corresponding dimensions to determine whether solids are similar. | 1  2  3  4 | |
| I can use corresponding dimensions to find missing measures in similar solids. | 1  2  3  4 | |
| I can use linear measures to find surface areas and volumes of similar solids. | 1  2  3  4 | |

Name _____ Date _____

**The solids are similar. Find the missing measure(s).**

**1.**

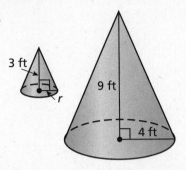

3 ft

9 ft

4 ft

r

**2.**

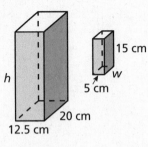

15 cm

h

5 cm

w

20 cm

12.5 cm

**The solids are similar. Find the surface area *S* or the volume *V* of the smaller solid. Round your answers to the nearest tenth.**

**3.** Surface Area = 294.7 m$^2$

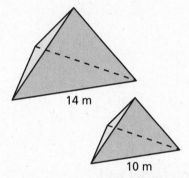

14 m

10 m

**4.** Volume = 1500 ft$^3$

6 ft

3.2 ft

**5.** The ratio of the corresponding linear measures of two similar buckets of popcorn is 2 to 5. The larger bucket has a volume of 390 cubic inches. Find the volume of the smaller bucket. Round your answer to the nearest tenth.

**6.** A box of 60 tissues has a length of 11 centimeters, a width of 10.5 centimeters, and a height of 13.5 centimeters.

   **a.** Find the volume of the box of tissues. Round your answer to the nearest tenth.

   **b.** A similar box contains 100 tissues. The ratio of the corresponding linear measures of the two boxes is 3 : 5. Find the volume of the larger box. Round your answer to the nearest tenth.

   **c.** Find the dimensions of the larger box. Round your answers to the nearest tenth.

## Chapter 16   Chapter Self-Assessment

**Use the scale below to rate your understanding of the learning target and the success criteria.**

| *1* | *2* | *3* | *4* |
|---|---|---|---|
| I do not understand. | I can do it with help. | I can do it on my own. | I can teach someone else. |

|  | Rating | Date |
|---|---|---|
| **16.1 Volumes of Cylinders** | | |
| **Learning Target:** Find the volume of a cylinder. | 1  2  3  4 | |
| I can use a formula to find the volume of a cylinder. | 1  2  3  4 | |
| I can use the formula for the volume of a cylinder to find a missing dimension. | 1  2  3  4 | |
| **16.2 Volumes of Cones** | | |
| **Learning Target:** Find the volume of a cone. | 1  2  3  4 | |
| I can use a formula to find the volume of a cone. | 1  2  3  4 | |
| I can use the formula for the volume of a cone to find a missing dimension. | 1  2  3  4 | |
| **16.3 Volumes of Spheres** | | |
| **Learning Target:** Find the volume of a sphere. | 1  2  3  4 | |
| I can use a formula to find the volume of a sphere. | 1  2  3  4 | |
| I can use the formula for the volume of a sphere to find the radius. | 1  2  3  4 | |
| I can find volumes of composite solids. | 1  2  3  4 | |

# Chapter 16 Chapter Self-Assessment (continued)

| | Rating | Date |
|---|---|---|
| **16.4 Surface Areas and Volumes of Similar Solids** | | |
| **Learning Target:** Find the surface areas and volumes of similar solids. | 1  2  3  4 | |
| I can use corresponding dimensions to determine whether solids are similar. | 1  2  3  4 | |
| I can use corresponding dimensions to find missing measures in similar solids. | 1  2  3  4 | |
| I can use linear measures to find surface areas and volumes of similar solids. | 1  2  3  4 | |

# Solving Multi-Step Equations
**For use with Exploration Topic 1**

**Learning Target:**   Write and solve multi-step equations.

**Success Criteria:**   • I can apply properties to produce equivalent equations.
   • I can solve multi-step equations.
   • I can use multi-step equations to model and solve real-life problems.

**1   EXPLORATION:** Finding Angle Measures

**Work with a partner. Find each angle measure in each figure. Use equations to justify your answers.**

a.
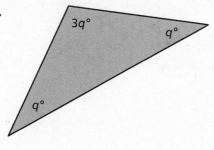

b.
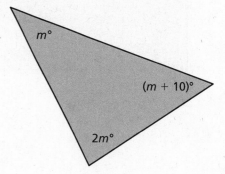

c.
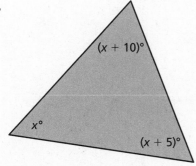

d.
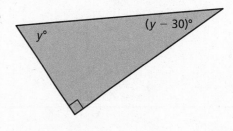

**Big Ideas Math: Modeling Real Life Grade 7 Accelerated**   **413**

**Topic 1** Solving Multi-Step Equations (continued)

e.

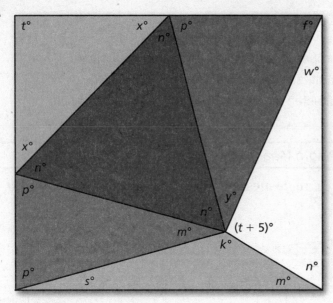

# Notetaking with Vocabulary

**Vocabulary:**

**Notes:**

# Self-Assessment

**Use the scale below to rate your understanding of the learning target and the success criteria.**

| 1 | 2 | 3 | 4 |
|---|---|---|---|
| I do not understand. | I can do it with help. | I can do it on my own. | I can teach someone else. |

|  | Rating | Date |
|---|---|---|
| **Topic 1 Solving Multi-Step Equations** | | |
| **Learning Target:** Write and solve multi-step equations. | 1    2    3    4 | |
| I can apply properties to produce equivalent equations. | 1    2    3    4 | |
| I can solve multi-step equations. | 1    2    3    4 | |
| I can use multi-step equations to model and solve real-life problems. | 1    2    3    4 | |

Name _____ Date _____

**Solve the equation. Check your solution.**

**1.** $\frac{5}{7}p - \frac{2}{7}p + 12 = 6$

**2.** $2.1x + 1.3x - 4.6 = 2.2$

**3.** $3(5 - 2h) + 9 = -30$

**4.** $14(x - 3) - 22x = -18$

**5.** The sum of the measures of the interior angles of the triangle is 180°. Write and solve an equation to find the value of the variable.

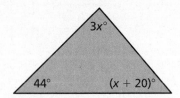

**6.** A rectangular field has an area of 2100 square feet. The length of the field is 50 feet.

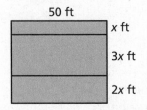

   **a.** How wide is the field?

   **b.** The field is divided into 3 rectangles, as shown. Write and solve an equation to find $x$.

   **c.** Determine the dimensions of each rectangle.

**7.** You are researching the price of MP3 players. You found an average price of $58.80. One MP3 player costs $56 and another costs $62. Find the price of the third MP3 player.

**8.** The perimeter of a triangle is 42 inches. One side measures 18 inches. The shortest side measures $x$ inches. The longest side measures 1 inch less than four times the length of the shortest side. Write and solve an equation to find the length of the longest side.

**9.** You order 4 fish sandwiches and a hamburger. The cost of the hamburger is $2.50. Your total bill before tax is $14.30. Write and solve an equation to find the cost of a fish sandwich.

**10.** A food service in the mall prepares free samples of chicken to give out during dinner time. One hour later, the food service has given out 4 fewer than 70% of the total number of samples. How many samples did the food service prepare if they gave out 80 samples in the first hour?

# Topic 2

## Solving Equations with Variables on Both Sides
**For use with Exploration Topic 2**

**Learning Target:** Write and solve equations with variables on both sides.

**Success Criteria:**
- I can explain how to solve an equation with variables on both sides.
- I can determine whether an equation has one solution, no solution, or infinitely many solutions.
- I can use equations with variables on both sides to model and solve real-life problems.

---

**1 EXPLORATION: Finding Missing Measures in Figures**

**Work with a partner.**

**a.** If possible, find the value of *x* so that the value of the perimeter (in feet) is equal to the value of the area (in square feet) for each figure. Use an equation to justify your answer.

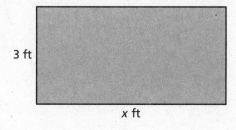

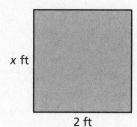

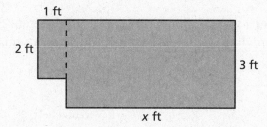

**Topic 2** Solving Equations with Variables on Both Sides (continued)

**b.** If possible, find the value of $y$ so that the value of the surface area (in square inches) is equal to the value of the volume (in cubic inches) for each figure. Use an equation to justify your answer.

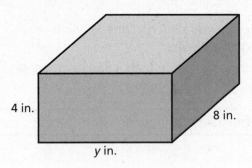

4 in.

8 in.

$y$ in.

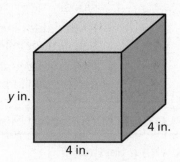

$y$ in.

4 in.

4 in.

**c.** How are the equations you used in parts (a) and (b) different from equations used in previous sections? Explain how to solve this type of equation.

# Notetaking with Vocabulary

**Vocabulary:**

**Notes:**

# Self-Assessment

**Use the scale below to rate your understanding of the learning target and the success criteria.**

| 1 | 2 | 3 | 4 |
|---|---|---|---|
| I do not understand. | I can do it with help. | I can do it on my own. | I can teach someone else. |

| | Rating | Date |
|---|---|---|
| **Topic 2 Solving Equations with Variables on Both Sides** | | |
| **Learning Target:** Write and solve equations with variables on both sides. | 1   2   3   4 | |
| I can explain how to solve an equation with variables on both sides. | 1   2   3   4 | |
| I can determine whether an equation has one solution, no solution, or infinitely many solutions. | 1   2   3   4 | |
| I can use equations with variables on both sides to model and solve real-life problems. | 1   2   3   4 | |

## Topic 2 Practice

**If possible, find the value of *x* so that the value of the surface area is equal to the value of the volume.**

**1.**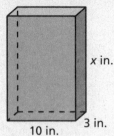

*x* in.

10 in. 3 in.

**2.**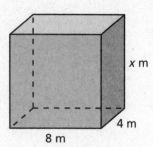

*x* m

4 m

8 m

**Solve the equation. Check your solution.**

**3.** $\frac{4}{7}m = 18 - \frac{2}{7}m$

**4.** $4(2s - 3) = 3(s + 1)$

**5.** Your cellular telephone provider offers two plans. Plan A has a monthly fee of $15 and $0.25 per text. Plan B has a monthly fee of $20 and $0.05 per text. Write and solve an equation to find the number of texts than you must send to have the same cost for each of the plans.

**6.** Describe and correct the error in solving the equation.

$$
\begin{aligned}
0.4x &= 0.2(x - 8) \\
0.4x &= 0.2x - 8 \\
0.2x &= -8 \\
x &= -4
\end{aligned}
$$

**Solve the equation. Check your solution, if possible.**

**7.** $4.2x - 3 = 0.5(8.4x + 6)$

**8.** $-\frac{1}{2}x + 1\frac{1}{2} = \frac{1}{2}(3 - x)$

**9.** The original price *p* for a necklace is the same at Store A and Store B. At Store A, the sale price is 60% of the original price. Last month, at Store B, the sale price was $40 less than the original price. This month, Store B is selling the necklace for 80% of last month's reduced price, making this month's sale price at Store B equal to the sale price at Store A. Write and solve an equation to find the original price of the necklace.

**10.** A yoga studio charges a $36 membership fee and $20.60 per month for 10 classes. A Martial Arts studio charges a $20 membership fee and $22.20 per month for 10 classes. Your friend belongs to the yoga studio the same month you belong to the Martial Arts studio. After how many months is your friend's total cost the same as your total cost?

## Topic 3  Rewriting Equations and Formulas
For use with Exploration Topic 3

**Learning Target:** Solve literal equations for given variables and convert temperatures.

**Success Criteria:**
- I can use properties of equality to rewrite literal equations.
- I can use a formula to convert temperatures.

---

**1  EXPLORATION:** Rewriting Formulas

**Work with a partner.**

**a.** Write a formula for the height $h$ of each figure. Explain your method.

- A parallelogram with area $A$ and base $b$.

- A rectangular prism with volume $V$, length $\ell$, and width $w$

- A triangle with area $A$ and base $b$

**b.** Write a formula for the length $\ell$ of each figure. Explain your method.

- A rectangle with perimeter $P$ and width $w$

- A rectangular prism with surface area $S$, width $w$, and height $h$

**Topic 3** **Rewriting Equations and Formulas** (continued)

c. Use your formulas in parts (a) and (b) to find the missing dimension of each figure.

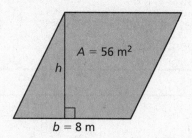

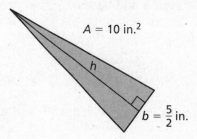

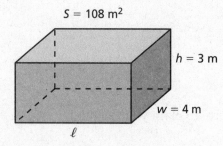

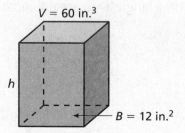

**Notetaking with Vocabulary**

**Vocabulary:**

**Notes:**

 **Self-Assessment**

Use the scale below to rate your understanding of the learning target and the success criteria.

| 1 | 2 | 3 | 4 |
|---|---|---|---|
| I do not understand. | I can do it with help. | I can do it on my own. | I can teach someone else. |

| | Rating | Date |
|---|---|---|
| **Topic 3 Rewriting Equations and Formulas** | | |
| **Learning Target:** Solve literal equations for given variables and convert temperatures. | 1   2   3   4 | |
| I can use properties of equality to rewrite literal equations. | 1   2   3   4 | |
| I can use a formula to convert temperatures. | 1   2   3   4 | |

## Topic 3 Practice

**Solve the equation for _y_.**

**1.** $3x - \frac{1}{4}y = -2$

**2.** $4.5x - 1.5y = 5.4$

**3.** The formula for the volume of a rectangular prism is $V = \ell wh$.

   **a.** Solve the formula for _w_.

   **b.** Use the new formula to find the value of _w_ when $V = 210$ cubic feet, $\ell = 10$ feet, and $h = 3$ feet.

**Solve the equation for the bold variable. Explain your method.**

**4.** $S = \pi r^2 + 2\pi r\boldsymbol{h}$

**5.** $A = \frac{1}{2}\boldsymbol{P}a$

**6.** The formula $F = \frac{9}{5}C + 32$ converts temperatures from Celsius _C_ to Fahrenheit _F_.

   **a.** Solve the formula for _C_.

   **b.** The boiling point of water is 212°F. What is the temperature in Celsius?

   **c.** If a house thermostat is set at 80°F, what is the setting in Celsius? Round your answer to the nearest tenth.

**7.** The formula for the area of a sector of a circle is $A = \dfrac{m}{360}\pi r^2$, given the measure _m_ of the angle and the radius _r_ of the circle.

   **a.** Solve the formula for _m_.

   **b.** Find the measure of the angle when the area of the sector is 5 square centimeters and the radius is 2 centimeters. Round your answer to the nearest tenth.

   **c.** If the area of the sector in part (b) is greater than 5 square centimeters, is the measure of the angle _greater than_ or _less than_ the answer to part (b)? Explain.

**8.** The formula for simple interest is $I = Prt$.

| $I$ | \$135 |
|-----|-------|
| $P$ |       |
| $r$ | 6% |
| $t$ | 3 years |

   **a.** Solve the formula for _P_, when _r_ is the simple interest per year.

   **b.** Use the new formula to find the value of _P_ in the table.

# Photo Credits

**111** andrearoad/iStock Unreleased/Getty Images Plus;
**162** ©iStockphoto.com/Joe Potato Photo;
**163** *top* Meral Hydaverdi/Shutterstock.com, *bottom*
Warren Goldswain/Shutterstock.com;
**351** *a.* Tom C Amon/Shutterstock.com,
*b.* Olga Gabay/Shutterstock.com;
*c.* Tom C Amon/Shutterstock.com;
*d.* HuHu/Shutterstock.com


**Cover Image** briddy_/iStock/Getty Images Plus

# Spinner 1

# Spinner 2

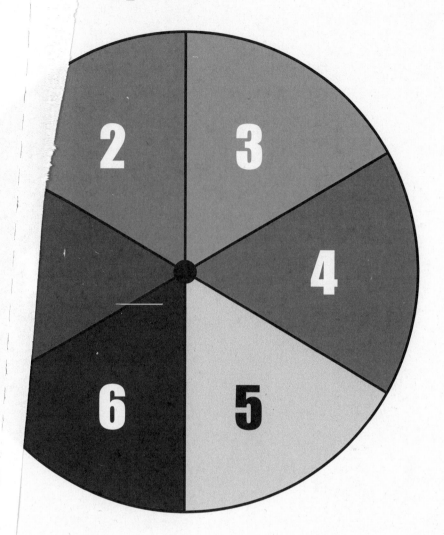

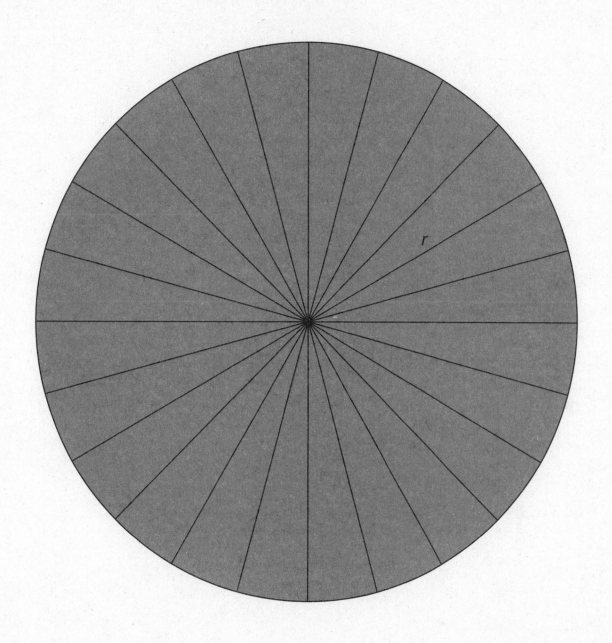

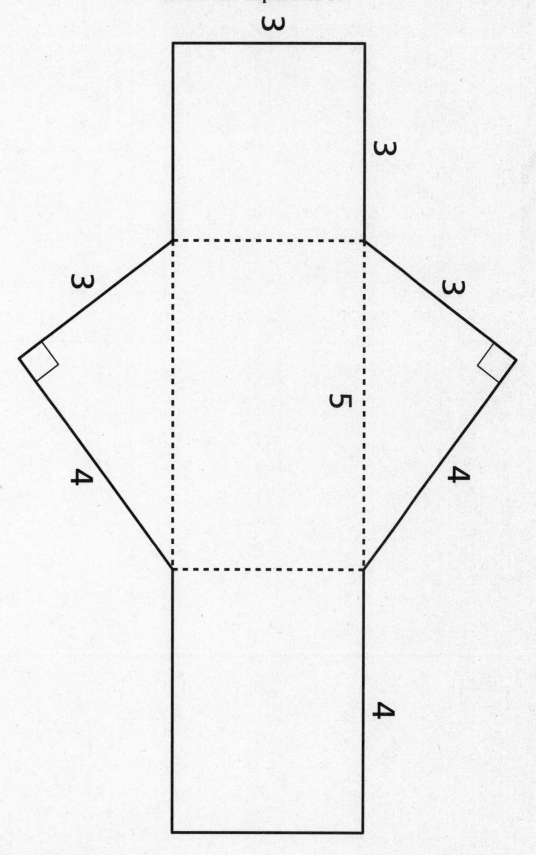

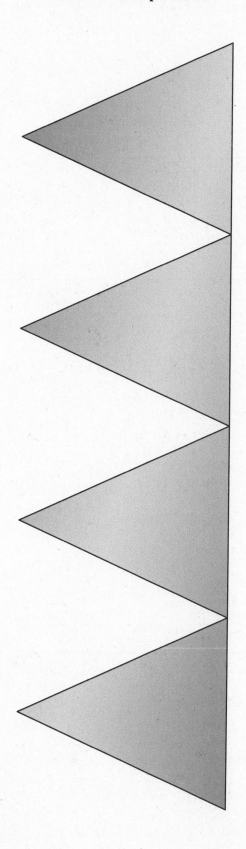

*Available at *BigIdeasMath.com.*

*Available at *BigIdeasMath.com.*

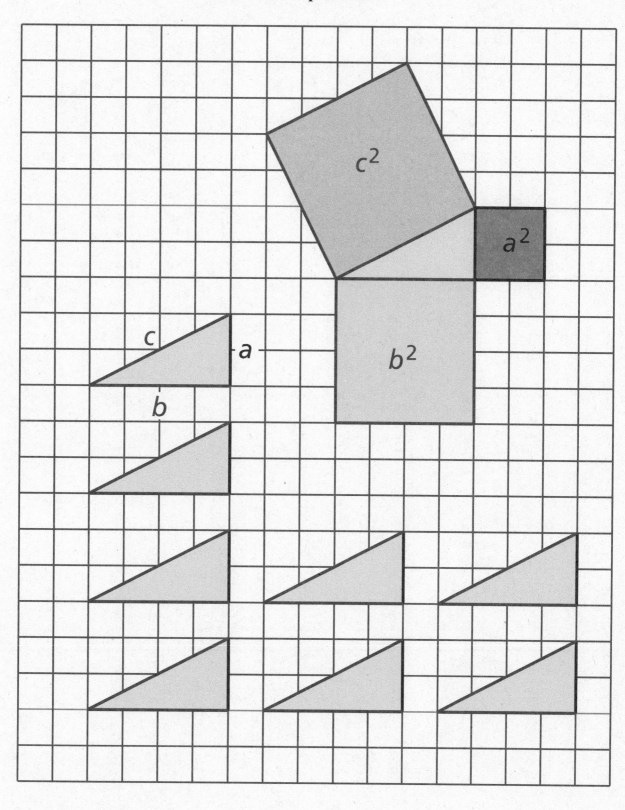

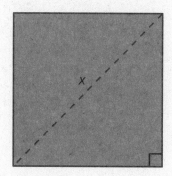

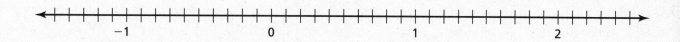

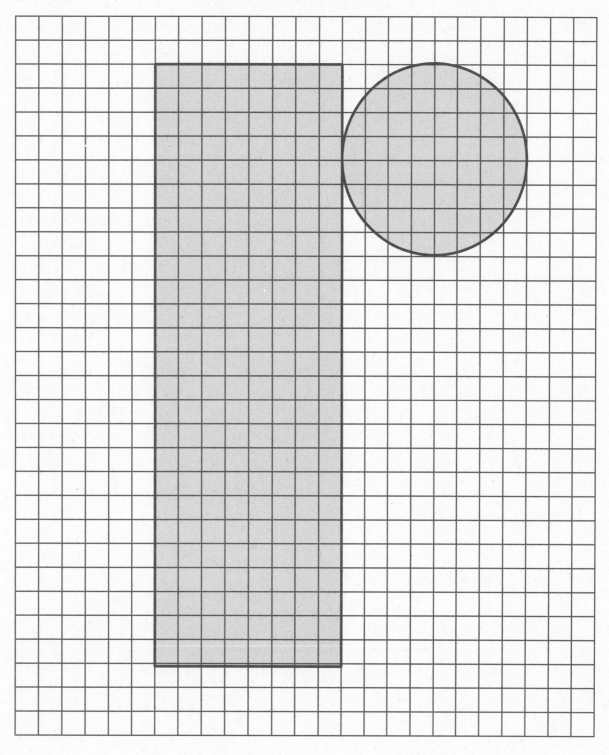

*Available at *BigIdeasMath.com.*

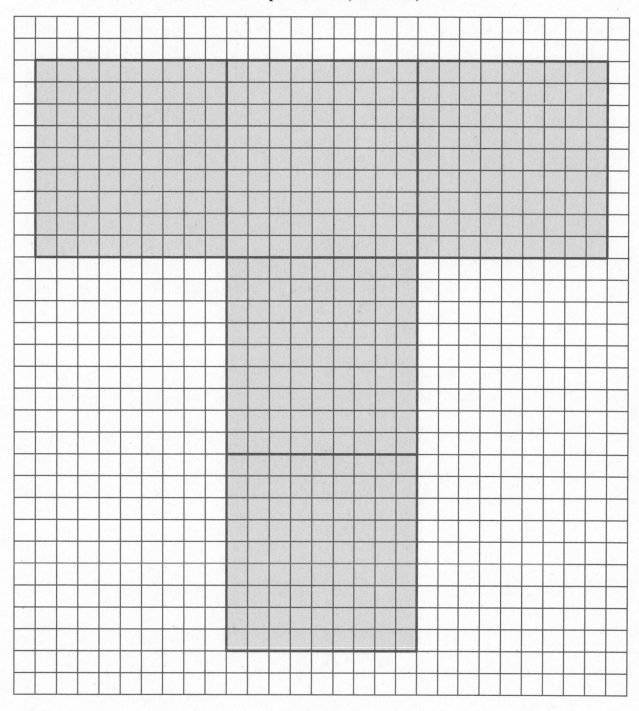

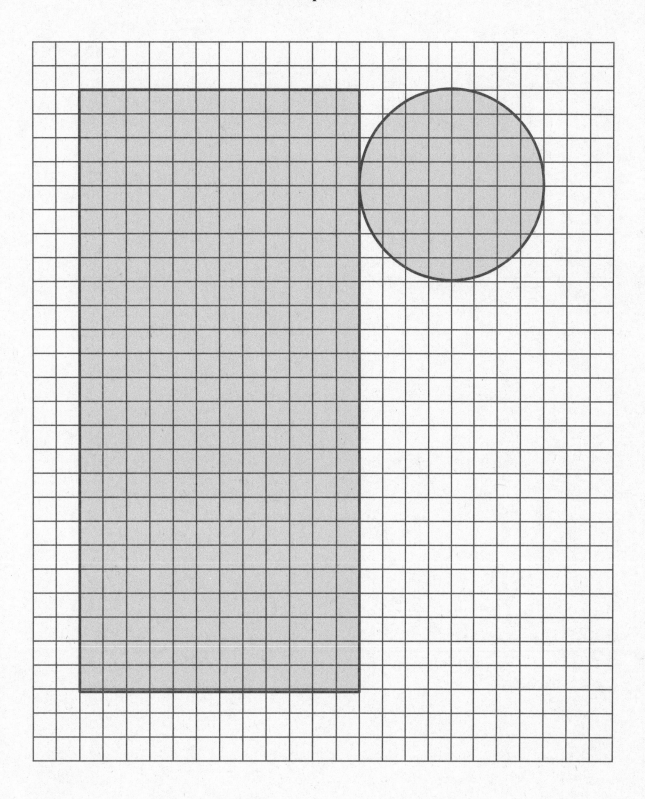

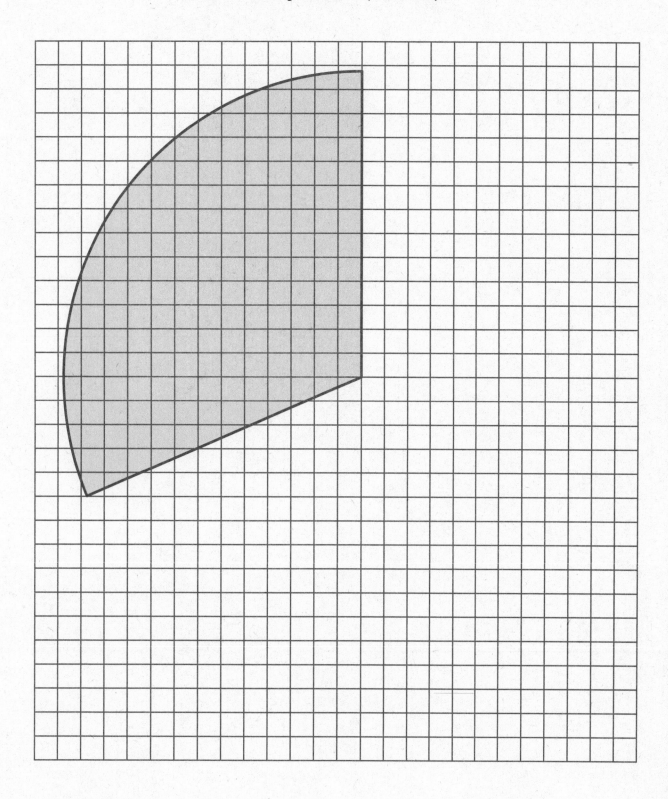

*Available at *BigIdeasMath.com*.

**Big Ideas Math: Modeling Real Life Grade 7 Accelerated**

# Integer Counters*

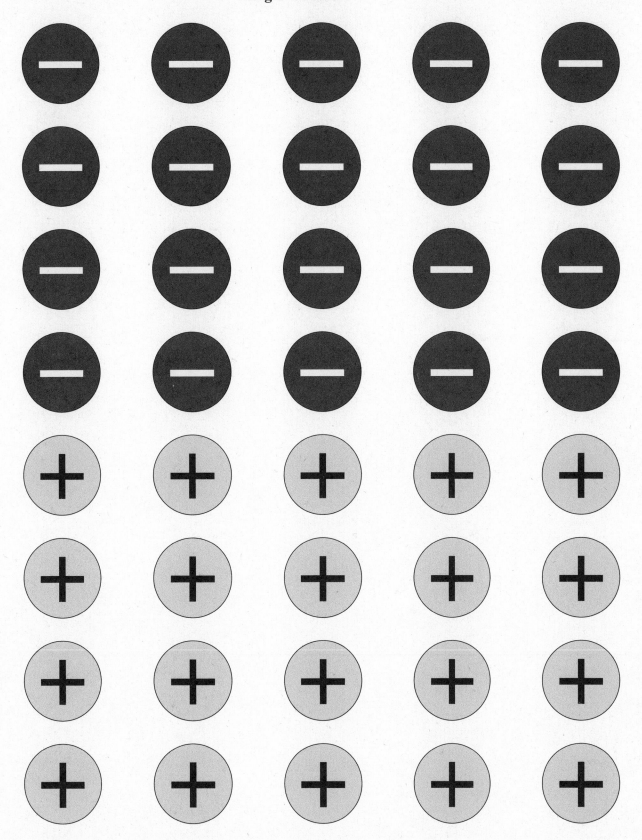

*Available at *BigIdeasMath.com*.

# Algebra Tiles*

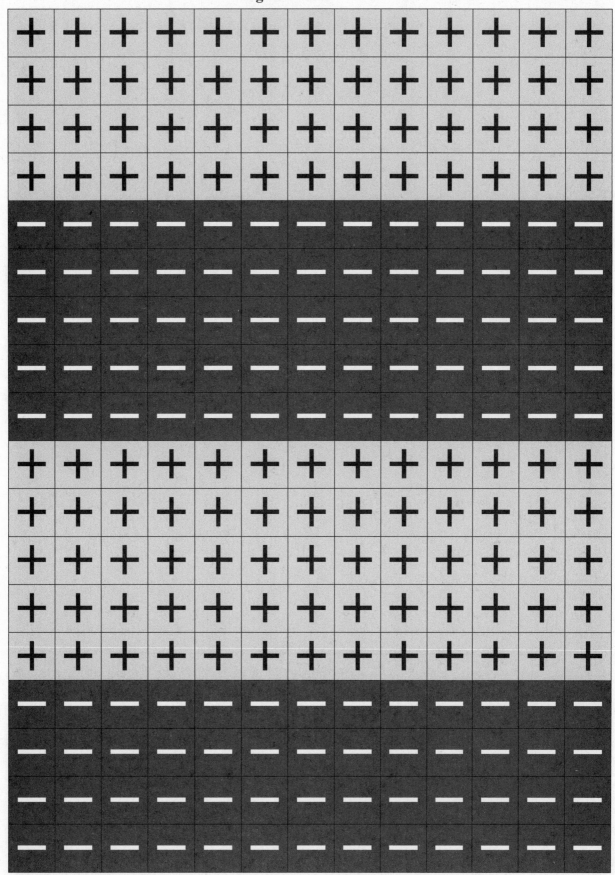

# Algebra Tiles (continued)*

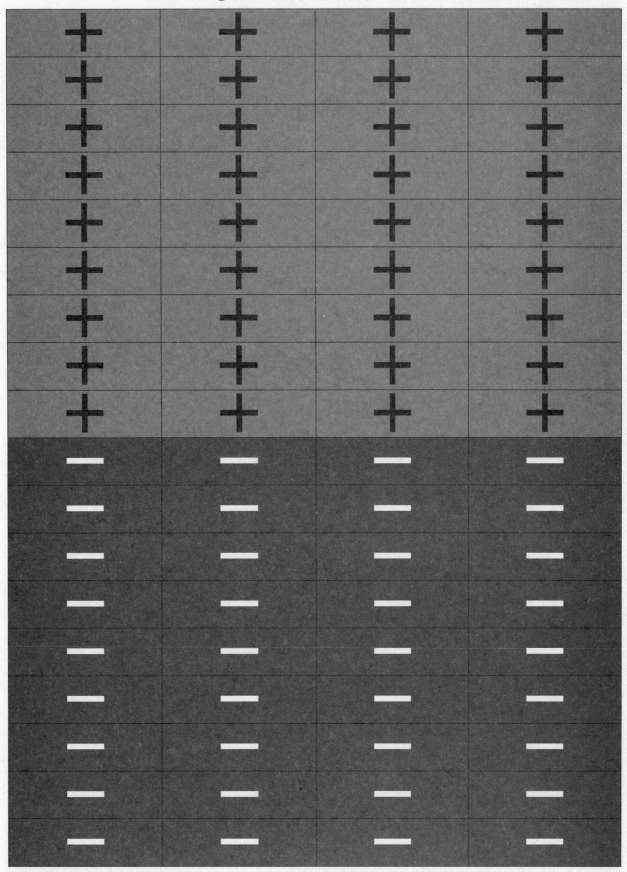

*Available at *BigIdeasMath.com*.